Plate Buckling in Bridges and other Structures

Plate Buckling in Bridges and other Structures

Björn Åkesson

Consulting Engineer, Fagersta, Sweden

Routledge
Taylor & Francis Group

LONDON AND NEW YORK

First published 2007 by Routledge

2 Park Square, Milton Park, Abingdon, Oxfordshire OX14 4RN
52 Vanderbilt Avenue, New York, NY 10017

*Routledge is an imprint of the Taylor & Francis Group, an informa
business*

First issued in paperback 2019

Typeset by Charon Tec Ltd (A Macmillan company), Chennai, India

British Library Cataloguing in Publication Data
A catalogue record for this book is available from the British Library

Library of Congress Cataloging in Publication Data
Åkesson, B. (Bjorn)
  Plate buckling in bridges and other structures / B. Åkesson.
    p. cm.
  Includes bibliographical references.
  ISBN 978-0-415-43195-8 (hardcover : alk. paper)
  1. Buckling (Mechanics) 2. Structural stability. I. Title.

TA656.2.A33 2007
624.1'76--dc22

                                  2006102580

ISBN13: 978-0-415-43195-8 (hbk)
ISBN13: 978-0-367-38945-1 (pbk)

Contents

Preface

As a lecturer at Chalmers University of Technology in Gothenburg, Sweden, during the years 1994–2004, I recognized the need for the students to have a pedagogical textbook concerning buckling of thin-walled plates. The books we used were often too theoretical – theory is essential, but it should be combined with practical issues as well. I therefore devoted my last two years at Chalmers to writing a textbook that would meet the needs of the students, and by extension, practising engineers. In writing the book, and delivering the information and disclosing the inner core of a complex subject, I tried to have in mind the learning process of the students.

Some may perhaps wonder – especially those readers looking for a book focusing exclusively on plane plates – why there is a chapter devoted to the buckling of shells? This final theoretical chapter ties together with the rest of the book, as there are important differences (and similarities) in the action of a shell in relation to a plane plate, which helps the reader to understand both the former and the latter. And one must also remember that even though Robert Stephenson's Britannia Bridge was built using only plane plates back in the 1850s, Stephenson, prior to the completion of the bridge, carried out tests on circular and elliptical girder tubes – one of the earliest examples of comparative tests to see the difference in buckling behaviour between different girder shapes.

January 2007
Björn Åkesson

Acknowledgement

During the ten-year-period in between 1994 and 2004 I was a lecturer and researcher within the fields of Structural Engineering and Bridge Engineering at Chalmers University of Technology in Göteborg, Sweden. My focus as a researcher was in the beginning concentrated on the fatigue life of riveted railway bridges in steel, and this was also reflected in the lectures I gave. However, the need and interest of the students did turn this focus more and more towards bridge engineering in general, and buckling of thin-walled plated bridge girders in particular. The one and only person that really did open my eyes and inspired me to gain deep knowledge within this field was Prof. Em. Bo Edlund. He has since the early 1970s been one of the leading researchers in the world concerning buckling problems of both thin-walled plated structures and cylinders. It has been a great privilege and honour working close to this extraordinarily talented man. Another good friend of mine, as well as research colleague, is Associate Prof. Mohammad Al-Emrani, with whom I spent numerous hours discussing different problems, mostly concerning fatigue, but also about buckling. Mohammad and Bo have been a great source of support during my years at Chalmers. I will also take the opportunity to thank Robert Kliger, the present professor at the department. I owe a lot to Robert, as it is entirely him, and no one else, who made it possible for me to write a textbook about buckling (an early, but short version of this book), during my last months at the department. Another good friend of mine, Jan Sandgren, has also contributed in the making of this book. He has over the years provided me with many ideas, articles and illustrations.

Björn Åkesson

List of symbols

α	reduction factor (production method, tolerance level)
β_A	the ratio between effective and gross area
γ_G	partial factor (load effect)
γ_{M0}	partial factor (resistance of cross-sections)
γ_{M1}	partial factor (resistance of members to instability)
η	knock-down factor
η	factor, taking the steel grade into account
$\bar{\eta}_1$	the ratio between design bending moment and design plastic resistance moment
$\bar{\eta}_3$	the ratio between design shear force and shear buckling design resistance
θ	slope of the panel diagonal
λ	slenderness parameter
$\bar{\lambda}_F$	slenderness parameter (concentrated load)
$\bar{\lambda}_p$	slenderness parameter (plate buckling)
λ_s	slenderness parameter (shell buckling)
$\bar{\lambda}_w$	slenderness parameter (shear buckling)
ν	poisson's ratio
ρ	reduction factor
σ_{bb}	strength of the tension field
σ_{cr}	critical buckling stress
σ_{cr}^r	critical buckling stress (reduced value)
σ_{Euler}	critical buckling stress according to the Euler theory
$\sigma_{f \cdot Ed}$	longitudinal stress in the flange
τ_{cr}	critical shear buckling stress
τ_r	shear load range (repeated loading)
φ	inclination of the tension field
χ	buckling reduction factor
χ_F	reduction factor (concentrated load)
ψ	stress value
ω	out-of-plane deflection
ω	axial deformation
ω_r	out-of-plane deflection (repeated loading)

ω_r	initial imperfection
ω_s	reduction factor (shell buckling)
A	cross-sectional area
A	constant
A_{eff}	effective net area
A_{gross}	gross area
a	plate length
a	distance between vertical stiffeners
a_c	buckling length
a/b	panel aspect ratio
b	plate width
b_e	effective width
b_{eff}	effective breadth
b_f	flange width
b_w	depth of the web
b/t	slenderness ratio
c	anchorage length in the flange
c/t	slenderness ratio
$M_{c.Rd}$	design bending moment resistance
M_{sd}	design bending moment
D	plate bending stiffness
D	weight
d	depth of the web
E	modulus of elasticity (Young's modulus)
F	design concentrated (transverse) load
F_{cr}	critical buckling load
F_{Ed}	design transverse force
F_{Rd}	design transverse force resistance
F_{sd}	design concentrated (transverse) load
f_{rd}	design strength resistance
f_y	yield strength
f_{yf}	yield strength of the flange
f_{yw}	yield strength of the web
G	weight
g	weight per meter
g	width of the tension field
h	depth of the web
h_w	depth of the web
I	second moment of area
I_{st}	second moment of area (stiffener)
i	radius of gyration
k	buckling coefficient
k_F	buckling coefficient (concentrated load)
k_τ	buckling coefficient (shear load)
L	span length

L_{eff}	effective length
L_{cr}	buckling length
l_c	buckling length
l_r	measure length (initial imperfection)
l_y	effective loaded length
M	bending moment
$M_{c \cdot Rd}$	design bending moment resistance
M_{Ed}	design bending moment
M_f	flange only bending moment
$M_{f \cdot Rd}$	flange only bending moment resistance
M_{pl}	plastic resistance moment
$M_{pl \cdot Rd}$	design plastic resistance moment
M_{sd}	design bending moment
m	number of half-sine waves (long. direction)
m_1	factor
m_2	factor
$N_{b \cdot Rd}$	design axial force resistance
$N_{b \cdot sd}$	design axial force
$N_{c \cdot sd}$	design axial force
$N_{c \cdot Rd}$	design axial force resistance
n	number of half-sine waves (transv. direction)
P	axial load
P_{cr}	critical buckling load (axial load)
P_{max}	maximum load-carrying capacity
P_{sd}	design load
Q	weight
q	evenly distributed load
q	evenly distributed load (design value)
q_{cr}	critical buckling load
R	anchored load
$R_{a \cdot Rd}$	design crippling resistance
$R_{y \cdot Rd}$	design crushing resistance
r/t	slenderness ratio
s_c	anchorage length (compression flange)
s_s	stiff bearing length
s_t	anchorage length (tension flange)
s_y	dispersion length
t	plate thickness
t_{eq}	equivalent plate thickness
t_f	flange thickness
t_w	web plate thickness
V	shear load
ΔV	critical shear buckling resistance
$V_{bb \cdot Rd}$	design shear buckling resistance

$V_{bf,Rd}$	design shear buckling resistance (flange contribution)
$V_{b,Rd}$	design shear buckling resistance
$V_{bw,Rd}$	design shear buckling resistance (web contribution)
$V_{bw.Rd}$	design shear buckling resistance
V_{Ed}	design shear force
V_{sd}	design shear force
W_{eff}	elastic section modulus (effective net section)
W_{el}	elastic section modulus
W_{pl}	plastic section modulus
$y_{n.a.}$	position of the neutral axis

Chapter 1

Introduction

Buckling is an instability phenomenon that can occur if a slender and thin-walled plate – plane or curved – is subjected to axial pressure (i.e. compression). At a certain given critical load the plate will buckle very sudden in the out-of-plane transverse direction. The compressive force could besides coming from pure axial compression, also be generated by bending moment, shear or local concentrated loads, or by a combination among these. If the structural element is compact, the load-carrying capacity is governed by the yield stress of the material, rather than buckling strength capacity. If instead the element is slender and/or thin-walled, the buckling strength is governed by the so-called slenderness ratio – the buckling length over the radius of gyration for global buckling of a column or a strut, or the loaded width over the thickness of the plate for local buckling. A special form of instability, that has to be considered with great care in design, is the combined global and local buckling risk of a slender and thin-walled axially loaded plated column – the capacity could here be much lower than the two buckling effects analyzed separately. In this book, however, we will only concentrate on the latter instability phenomenon, i.e. local buckling.

Eurocode defines four cross-section classes with reference to the local buckling risk. The parameter that governs what particular class a cross-section belongs to is the slenderness ratio of the individual plates of the cross-section mentioned above. The level of the slenderness ratio then governs the ability (or inability) for plastic rotational capacity, i.e. elongation at the tension side, and compression (with possible buckling risk) at the other side, for a girder subjected to a bending moment. These four classes in the Eurocode (Class 1–4) are for girders subjected to a bending moment defined as follows below. The maximum possible loading capacity (q_{sd}) in the ultimate loading state is given by this condition (where index c is telling us that it is the ability to carry compressive stresses – with respect to the local buckling risk – that governs the maximum capacity) (Eq. 1.1):

$$M_{sd}^{max} \leq M_{c.Rd} \tag{1.1}$$

1.1 Class 1

The cross-section is so compact (read: with a sufficiently low slenderness ratio/high plastic rotational capacity) that it is possible to form a mechanism with plastic hinges in a statically in-determinate system. This gives the possibility to level out the bending moment differences in ultimate limit state design. Girders in class 1 are normally standard hot-rolled profiles (Fig. 1.1).

1.2 Class 2

The cross-section is also here compact, but not enough to be able to form a mechanism in a statically indeterminate system. The design moment distribution is for the elastic response. However, the longitudinal section with the maximum moment can be designed for full plastification over the entire cross-section height – in this case similar to class 1 cross-sections. For statically determinate systems there is no difference between the two classes. Girders in class 2 are normally also standard hot-rolled profiles (Fig. 1.2).

1.3 Class 3

The cross-section can be characterized as semi-compact, having a reduced capacity for full plastification, due to the local buckling risk on the compression side. Just as for class 2 profiles, the design moment distribution is for elastic response, however, with the difference that the maximum strained section is designed for elastic (triangular) stress distribution. Girders in class 3 are normally welded profiles (Fig. 1.3).

For unsymmetrical cross-sections (in class 3) – e.g. having a wider compression flange (than the tension flange) – yielding is accepted for the tensile stresses, however, the stresses at the compressive side limited to the yield strength at the extreme fibre. Where only the web is in class 3, and the compression flange is either in class 1 or 2, the Eurocode accepts that the properties are based on an effective class 2 cross-section, where complete yielding of the entire cross-section is accepted, with the exception of a central part of the web subjected to compression, which is neglected.

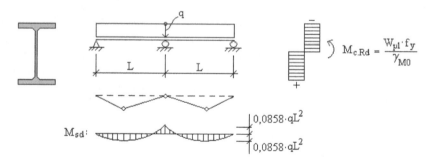

Figure 1.1 Cross-section class 1.

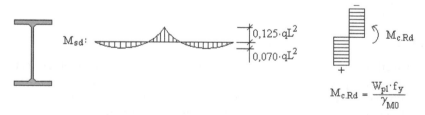

Figure 1.2 Cross-section class 2.

1.4   Class 4

The cross-section is thin-walled, i.e. having such a high slenderness ratio that buckling will occur before yielding is reached in the outermost fibre. Post-critical reserve effects enables though for yielding to be reached in the extreme fibre in the ultimate limit state design. An effective net cross-section is analyzed, where the buckled zone is removed from the gross cross-section (due to the loss of stiffness in that area). Examples of profiles in class 4 are welded bridge girders (Fig. 1.4).

For columns and struts, subjected to pure axial compression, the load-carrying capacity for profiles in class 1–3 is only reduced with respect to global (Euler) buckling risk. It is first at profiles in class 4 that the load-carrying capacity also has to be reduced for the local (plate) buckling risk.

In bridge construction, as well as in aircraft and shipbuilding industry, it is an absolute necessity to save material, and therefore the structural elements are made thin-walled and slender. To choose a compact profile (which is able to fully plastify before any local buckling risk) is not economical, as it wastes material – the increase in load-carrying capacity (in comparison to a thin-walled cross-section) is eaten up by the relative increase in cross-sectional area (compare example 9 in this book). In addition, it is absolutely necessary to keep the self-weight down, so that a good and sufficient part of the load-carrying capacity is spared for the traffic load (read: too much part of the load-carrying capacity should not be taken by the self-weight alone). A heavy and compact section bridge is also costly with respect to the extra need of foundation and substructure dimensions. High and slender girders (with thin-walled cross-sections)

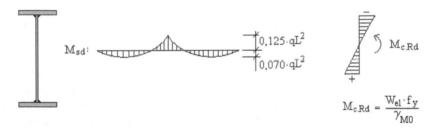

Figure 1.3   Cross-section class 3.

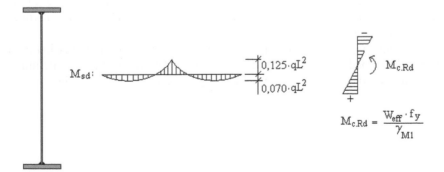

Figure 1.4   Cross-section class 4.

have also a higher stiffness relative an equivalent girder having a compact cross-section, leading to a reduced deflection under loading. In construction of buildings though, compact sections (such as standard hot-rolled profiles) are preferred, as they keep the profile depth down. These compact profiles are also more "robust", which can be needed during transport, handling and assembly.

One way of further increasing the load-carrying capacity of a slender and thin-walled plate is by the help of stiffeners, which minimize the free spacing of the parts subjected to compression. A plated bridge girder, as well as the hull of a ship, is normally stiffened in both the longitudinal and the transverse direction in order to maximize the load-carrying capacity. Provided that the stiffeners are sufficiently strong, the risk of buckling is restricted to the plate areas in between the stiffeners. The maximum load-carrying capacity of these plate panels is then governed by the plate buckling risk, however, also by taking the post-critical reserve effects into account.

The general expression for the critical buckling stress (irrespective of the type of stress distribution) is (Eq. 1.2):

$$\sigma_{cr} = k \cdot \frac{\pi^2 \cdot E}{12 \cdot (1 - \upsilon^2) \cdot \left(\dfrac{b}{t}\right)^2} \tag{1.2}$$

The so-called buckling coefficient k varies depending on the type of stress distribution, and on the quotient between the length (denoted a) and the width (denoted b) of the plate (k has its lowest value for pure axial loading in compression, which also gives the lowest value for the critical buckling stress). The quotient b/t is the slenderness (ratio) of the plate.

Plate buckling has – in contrary to global buckling of a column or a strut, or the lateral-torsional buckling of a beam – a post-critical load-carrying capacity that enables for additional loading after local buckling has occurred. A plate is in that sense inner statically indeterminate, which makes the collapse of the plate not coming when buckling occurs, but instead later, at a higher loading level. This is taken into consideration in the ultimate limit state design of plates – local buckling does not restrict the load-carrying capacity to the critical buckling stress, instead the maximum capacity consists of the two parts; the buckling load + the additional post-critical load. Global buckling of a column or a strut does not exhibit such an indeterminate behaviour, as these are statically determinate systems (having no post-critical reserve strength, i.e. no ability to redistribute load). This particular instability phenomenon – global buckling of a strut or a column – is, however, not the focus of this book.

In the coming chapter we will concentrate more in detail on the theory behind plate buckling and the load-carrying capacity of unstiffened plates in the ultimate limit state (for thin-walled cross-sections in class 4 as they are defined in Eurocode).

Chapter 2

Plate buckling theory

Consider the axially loaded "plate strut" in Fig. 2.1 (width b, length a, and thickness t), having the loaded edges supported and the unloaded edges free. The strut has the appearance of a plate, however, but not treated as such – we will instead use the classical Euler theory in our following analysis (and soon come to the theory for true plate action).

When the load reaches a certain critical value, expressed as either P_{cr} or σ_{cr}, the strut buckles and collapses (read: the lateral deflection goes to infinity) (Fig. 2.2).

For any given axial loading below this critical value, it is possible to apply an additional horizontal (transverse) force without the occurrence of buckling (the strut balances both the vertical – axial – loading and the horizontal, and will deflect back

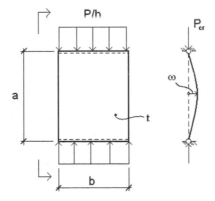

Figure 2.1 An axially loaded "plate strut".

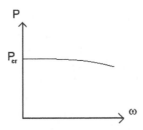

Figure 2.2 Load/displacement curve for an axially loaded "plate strut".

as soon as the horizontal load is removed). The closer the axial load is to the critical buckling load, the less the ability to carry an additional horizontal loading becomes. At exactly the critical buckling load, this ability becomes zero – the strut is then barely able to just carry the axial load.

According to the well-known Euler theory, the critical buckling load for the strut becomes (Eq. 2.1):

$$P_{cr} = \frac{\pi^2 \cdot EI}{a^2} \cdot \frac{1}{(1 - v^2)} \tag{2.1}$$

This expression is adjusted with respect to the relatively large width in relation to the (buckling) length of the strut that we are studying. This adjustment is done with the quotient $1/(1 - v^2)$, and this is due to the free strain deformations in the transverse direction in the centre part, in relation to the constraint at the loaded edges. In comparison to the normal appearance of a strut – where the width is small in comparison to the length – we will, for our plate strut, receive a slightly higher value for the critical buckling load due to this transverse strain divergence.

We now transform the critical buckling load (P_{cr}) given above, to an equivalent critical buckling stress (σ_{cr}), with the help of the expression for the moment of inertia of the strut (Eq. 2.2):

$$\sigma_{cr} = \frac{P_{cr}}{b \cdot t} \Rightarrow \quad \sigma_{cr} = \frac{\pi^2 \cdot E}{12 \cdot (1 - v^2) \cdot \left(\frac{a}{t}\right)^2} \tag{2.2}$$

$$I = \frac{b \cdot t^3}{12}$$

Now compare the small difference between the expressions for the critical buckling stress of a plate (see Eq. 1.2) with the expression above for the (Euler) strut. If one remembers how easily this latter expression was found, then it also works as a good reminder of the former! The critical buckling stress expression for a plate has a factor k (called the buckling coefficient), and a slenderness ratio defined as the quotient b/t, instead of a/t for the strut, otherwise the expressions are similar. The last difference regarding the definition of the slenderness – that it is the width b instead of the length a over the thickness t – is very important to remember, as the length of a plate is not governing the critical buckling stress. The width b is the main parameter governing the critical buckling stress of a plate, and we will next find out – by the help of the theory behind plate buckling – why this is so.

By definition, a strut (or a column) is only supported at its loaded edges, while a plate is supported at three edges or more, and it is this fact that makes a plate have a different buckling behaviour than a strut – the transverse width b becomes the governing parameter instead of the length a. It is also in the transverse direction relative the loading direction, that plates have a capacity to develop a tension field after buckling has occurred, and by doing so – through a transverse membrane action – enable for an additional loading capacity in the so-called post-critical range.

In order to get a background to the expression for the critical buckling stress of a plate – that was given in chapter 1 (Eq. 1.2) – we start by studying the differential equation for a plate, however, not loaded in the axial direction, but in the *transverse* direction by bending (Fig. 2.3).

The differential equation gives us the relationship between the lateral out-of-plane deflection and the transverse loading q (Eq. 2.3):

$$D \cdot \left(\frac{\delta^4 \omega}{\delta x^4} + 2 \cdot \frac{\delta^4 \omega}{\delta x^2 \delta y^2} + \frac{\delta^4 \omega}{\delta y^4} \right) = q \tag{2.3}$$

where D is the plate bending stiffness (Eq. 2.4):

$$D = \frac{E \cdot t^3}{12 \cdot (1 - v^2)} \quad [\text{Nm}^2/\text{m}] \tag{2.4}$$

Compare the differential equation given above (in Eq. 2.3), with the equivalent 2D-expression for a beam also subjected to bending (Fig. 2.4) (Eqs. 2.5–2.8):

$$-EI \cdot \frac{\delta^2 \omega}{\delta x^2} = M \tag{2.5}$$

$$M(x) = \frac{q \cdot L}{2} \cdot x - \frac{q \cdot x^2}{2} \tag{2.6}$$

$$M'(x) = \frac{q \cdot L}{2} - q \cdot x \tag{2.7}$$

$$M''(x) = -q \quad \Rightarrow \quad EI \cdot \frac{\delta^4 \omega}{\delta x^4} = q \tag{2.8}$$

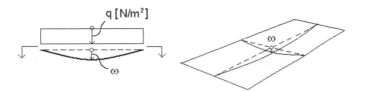

Figure 2.3 A plate loaded in the transverse direction by an evenly distributed load, q.

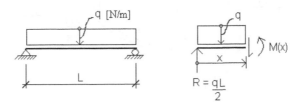

Figure 2.4 A beam subjected to bending.

Let us now continue to study a plate, however, not loaded now in bending, but instead having an in-plane axial load (Fig. 2.5).

We look for the equilibrium of a small element (having $\delta x = \delta y = 1$) in a deflected state, axially loaded with a normal force per unit width, $\sigma_x \cdot t$ [N/m] (Fig. 2.6).

This small element does not have a load in the transverse direction, i.e. $q = 0$ when we compare with the differential equation (given in Eq. 2.3). It is true that for normal forces below the critical buckling load (i.e. in the *sub-critical* range) it is required an additional transverse load/force to keep the plate in a deflected shape, however, this deflection would go back as soon as this additional load would be removed. At a certain level of the axial load ($\sigma_x = \sigma_{cr}$), the outwards going and resulting transverse force – due to the curvature – is in precise balance with the "re-bouncing" force. This exact value of the normal force (or stress) is defined as the critical value with respect to plate buckling.

The differential equation for an axially loaded plate can thus be written (Eq. 2.9):

$$D \cdot \left(\frac{\delta^4 \omega}{\delta x^4} + 2 \cdot \frac{\delta^4 \omega}{\delta x^2 \delta y^2} + \frac{\delta^4 \omega}{\delta y^4} \right) = -\sigma_{cr} \cdot t \cdot \frac{\delta^2 \omega}{\delta x^2} \tag{2.9}$$

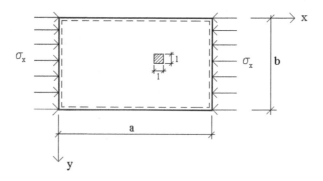

Figure 2.5 A plate loaded in the axial direction with an evenly distributed edge load.

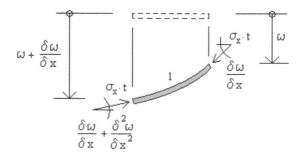

Figure 2.6 The deflected state of an axially loaded small element.

The general expression for the solution of this equation is as follows (where we have assumed a sine wave in both the longitudinal and transverse direction) (Eq. 2.10):

$$\omega = A \cdot \sin \frac{m \cdot \pi \cdot x}{a} \cdot \sin \frac{n \cdot \pi \cdot y}{b} \qquad (2.10)$$

ω out-of-plane deflection
A constant
m number of half-sine waves in the longitudinal direction
n number of half-sine waves in the transverse direction
a length of the plate (unloaded edge)
b width of the plate (loaded edge)

For $n=1$ and $m=2$ the buckling mode looks as follows (Fig. 2.7).

We use the general expression for the solution of the out-of-plane deflection, in order to find the value of the critical buckling stress, σ_{cr} (Eqs. 2.11–2.17):

$$\frac{\delta \omega}{\delta x} = A \cdot \frac{m\pi}{a} \cdot \cos \frac{m\pi x}{a} \cdot \sin \frac{n\pi y}{b} \qquad (2.11)$$

$$\frac{\delta^2 \omega}{\delta x^2} = -A \cdot \frac{m^2 \pi^2}{a^2} \cdot \sin \frac{m\pi x}{a} \cdot \sin \frac{n\pi y}{b} \qquad (2.12)$$

$$\frac{\delta^3 \omega}{\delta x^3} = -A \cdot \frac{m^3 \pi^3}{a^3} \cdot \cos \frac{m\pi x}{a} \cdot \sin \frac{n\pi y}{b} \qquad (2.13)$$

$$\frac{\delta^4 \omega}{\delta x^4} = A \cdot \frac{m^4 \pi^4}{a^4} \cdot \sin \frac{m\pi x}{a} \cdot \sin \frac{n\pi y}{b} \qquad (2.14)$$

$$\frac{\delta^3 \omega}{\delta x^2 \delta y} = -A \cdot \frac{m^2 \pi^2}{a^2} \cdot \sin \frac{m\pi x}{a} \cdot \frac{n\pi}{b} \cdot \cos \frac{n\pi y}{b} \qquad (2.15)$$

$$\frac{\delta^4 \omega}{\delta x^2 \delta y^2} = A \cdot \frac{m^2 \pi^2}{a^2} \cdot \sin \frac{m\pi x}{a} \cdot \frac{n^2 \pi^2}{b^2} \cdot \sin \frac{n\pi y}{b} \qquad (2.16)$$

$$\frac{\delta^4 \omega}{\delta y^4} = A \cdot \sin \frac{m\pi x}{a} \cdot \frac{n^4 \pi^4}{b^4} \cdot \sin \frac{n\pi y}{b} \qquad (2.17)$$

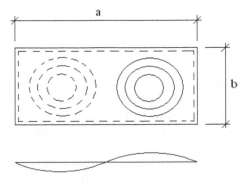

Figure 2.7 The buckling mode of a rectangular plate, having the relationship $a/b = 2$.

The differential equation has now become (Eqs. 2.18–2.20):

$$D \cdot \left(\frac{m^4 \pi^4}{a^4} + 2 \cdot \frac{m^2 \pi^2}{a^2} \cdot \frac{n^2 \pi^2}{b^2} + \frac{n^4 \pi^4}{b^4} \right) = \sigma_{cr} \cdot t \cdot \frac{m^2 \cdot \pi^2}{a^2} \tag{2.18}$$

$$\pi^4 \cdot D \cdot \left(\frac{m^2}{a^2} + \frac{n^2}{b^2} \right)^2 = \sigma_{cr} \cdot t \cdot \frac{m^2 \pi^2}{a^2} \tag{2.19}$$

$$\Rightarrow \quad \sigma_{cr} = \frac{\pi^2 \cdot D \cdot a^2}{t \cdot m^2} \cdot \left(\frac{m^2}{a^2} + \frac{n^2}{b^2} \right)^2 \tag{2.20}$$

The lowest value of σ_{cr} is received for $n = 1$, i.e. when there is only *one* half-sine wave in the transverse direction (Eq. 2.21):

$$\sigma_{cr} = \frac{\pi^2 \cdot D \cdot a^2}{t \cdot m^2} \cdot \left(\frac{m^2}{a^2} + \frac{1}{b^2} \right)^2 = \frac{\pi^2 \cdot D}{t} \cdot \frac{a^2}{m^2} \cdot \left(\frac{m^4}{a^4} + 2 \cdot \frac{m^2}{a^2 \cdot b^2} + \frac{1}{b^4} \right)$$

$$= \frac{\pi^2 \cdot D}{t} \cdot \left(\frac{m^2}{a^2} + \frac{2}{b^2} + \frac{a^2}{m^2} \cdot \frac{1}{b^4} \right) = \frac{\pi^2 \cdot D}{t} \cdot \frac{1}{b^2} \left(\frac{m^2 \cdot b^2}{a^2} + 2 + \frac{a^2}{m^2 \cdot b^2} \right)$$

$$= \frac{\pi^2 \cdot D}{t} \cdot \frac{1}{b^2} \cdot \left(\frac{m \cdot b}{a} + \frac{a}{m \cdot b} \right)^2 = k \cdot \frac{\pi^2 \cdot E}{12 \cdot (1 - v^2) \cdot \left(\frac{b}{t} \right)^2} \tag{2.21}$$

We have then also used the expression for the plate bending stiffness D (Eq. 2.4) and the so-called buckling coefficient k (as it is defined) (Eq. 2.22):

$$k = \left(\frac{m \cdot b}{a} + \frac{a}{m \cdot b} \right)^2 \tag{2.22}$$

The buckling coefficient k (or the "plate factor", as it is sometimes also called) is a function of the panel aspect ratio a/b, and the number of half-sine waves m in the longitudinal direction (i.e. in the loading direction). This coefficient has a minimum value of 4.0 for a given value of m and the same number for the quotient a/b (Fig. 2.8).

As for a rectangular plate, having a panel aspect ratio $a/b = 3$, the buckling mode that will give the lowest value of the critical buckling stress (with $k = 4.0$) will be the case where the number of half-sine buckling waves in the longitudinal direction divides the plate into the three unit squares, having an equally large buckle in each (i.e. $m = 3$) (Fig. 2.9).

A higher or less number of buckles would require more energy, and thus give a higher value of the critical buckling stress (sometimes give a higher value of the buckling coefficient k). Irrespective the size of the plate, it has in practical design been accepted

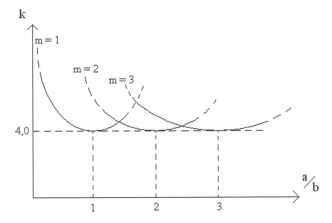

Figure 2.8 The buckling coefficient, k, as a function of the number of half-sine waves in the longitudinal direction, m, and the panel aspect ratio, a/b.

Figure 2.9 The buckling mode of a rectangular plate, having a panel aspect ratio a/b = 3.

the asymptotic value for the buckling coefficient (i.e. $k = 4.0$ for every case) – a value that is always on the safe side.

In comparison to an "Euler strut", the critical buckling stress for a plate does not decrease as the length a increases (k_{min} will always be 4.0) – instead it is the width b that governs the magnitude, as we did see for the solution of the differential equation. Given a fixed value for the width b (and a panel aspect ratio $a/b = 1$), the difference in critical buckling stress between a plate and a strut (having the same dimensions) will then be equal to 4, i.e. equal to the value of the buckling coefficient in the former case. This difference will then increase as the length a increases (the critical buckling stress of the plate will be constant, however, for the strut the same will decrease). If instead the length a is kept constant, and we gradually do increase the width b, the difference will then gradually diminish, and be close to identical for very wide plates. This is due to the fact that the curvature in the transverse direction is very small for wide plates, i.e. the central part can more or less be compared to a deflected strut having a simple curvature.

The discussion so far has been concentrated on plates, axially loaded and simply supported on all four edges – but what about other loading and boundary conditions? Except that an edge could be simply supported, it can also be fixed or being absolutely free (i.e. having no support at all). The stress distribution could also vary, besides being evenly distributed. It could for example be triangularly distributed (i.e. be the result

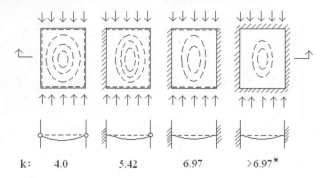

Figure 2.10 Evenly distributed axial compression – all four edges supported. The buckling coefficient, k, as a function of different edge conditions.
* Well above 6.97 for smaller panel aspect ratios, but close to this value for a/b ≥ 3.

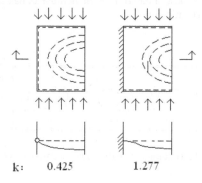

Figure 2.11 Evenly distributed load – only three edges supported. The buckling coefficient, k, as a function of different edge conditions.

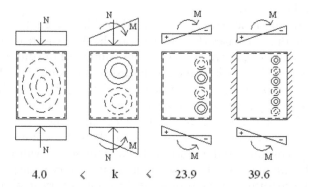

Figure 2.12 Different loading conditions – all four edges supported. The buckling coefficient, k, as a function of different edge conditions.

in-plane bending), or coming from the effect of shear along the edges). Both the edge conditions and the stress distribution affect the value of the buckling coefficient k. In the following Figs. 2.10–2.13, the minimum value of the buckling coefficient is given for different edge and loading conditions (Eqs. 2.23–2.25 for shear panels).

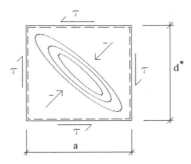

Figure 2.13 A plate panel subjected to shear. The buckling coefficient, k_τ, varies according to the value of the panel aspect ratio, *a/d*.
* The plate width is here designated *d* instead of *b*. In the Eurocode it is designated h.

$$\tau_{cr} = k_\tau \cdot \frac{\pi^2 \cdot E}{12 \cdot (1 - \upsilon^2) \cdot \left(\frac{d}{t}\right)^2} \qquad (2.23)$$

$$k_\tau = 5,34 + \frac{4}{\left(\frac{a}{d}\right)^2} \quad \left[\frac{a}{d} \geq 1\right] \qquad (2.24)$$

$$k_\tau = 4 + \frac{5,34}{\left(\frac{a}{d}\right)^2} \quad \left[\frac{a}{d} \leq 1\right] \qquad (2.25)$$

In this last loading case it should be noted that shear forces in a girder subjected to bending is also accompanied with bending moments, which has to be taken into consideration (see more in section 4.4).

The discussion so far has been about the critical buckling stress for plates, however, we shall in the following go more into detail why the ultimate loading capacity is not restricted to the occurrence of elastic buckling. As has been mentioned before, plates do possess ability for a post-critical reserve strength, which enables for an additional loading capacity after that buckling has occurred. This post-critical reserve strength is shown below in the load-/displacement diagram (Fig. 2.14).

As can be noted in the graph, the plate does not collapse at the so-called bifurcation point (read: the load/stress where the plate buckles out in any of the two transverse directions) as Euler struts do. Instead, the plate is able to carry additional load after buckling has occurred, and this is due to the formation of a membrane that stabilizes the buckle through a transverse tension band. When the central part of the plate buckles, it looses the major part of its stiffness, and then the load is forced to be "linked" around this weakened zone into the stiffer parts on either side. And due to this redistribution a transverse membrane in tension is formed and anchored. Study the plate below in the post-critical range, where the load paths is shown by the help of a strut-and-tie analogy (Fig. 2.15).

The action of a plate – in contrast to a strut – show an inherent statically indeterminacy, that enables for this ability to redistribute load after buckling has occurred.

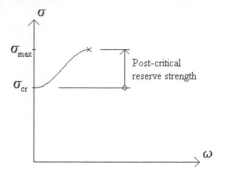

Figure 2.14 Stress/displacement diagram in the post-critical range.

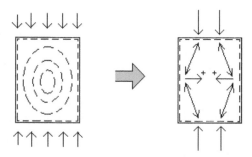

Figure 2.15 The redistribution of the transfer of load in the ultimate limit state (the post-critical range).

The maximum load-carrying capacity is governed by buckling of the stiffer edge zones as they have reached yielding (f_y), which was suggested by von Kármán in 1932. We study a plate in the post-critical range, where the stiffer edge zones will give us an effective width b_e (Fig. 2.16).

In order to obtain the maximum load-carrying capacity P_{max}, and the effective width b_e, we use the expression for the critical buckling stress and set it equal to the yield stress (Eqs. 2.26–2.28):

$$\sigma_{cr} = f_y = k \cdot \frac{\pi^2 \cdot E}{12 \cdot (1 - \upsilon^2) \cdot \left(\frac{b_e}{t}\right)^2} \tag{2.26}$$

$$b_e = \sqrt{\frac{k \cdot \pi^2}{12 \cdot (1 - \upsilon^2)}} \cdot \sqrt{\frac{E}{f_y}} \cdot t = 1,90 \cdot \sqrt{\frac{E}{f_y}} \cdot t \tag{2.27}$$

($k = 4$ and $\upsilon = 0,3$)

$$\Rightarrow \quad P_{max} = f_y \cdot b_e \cdot t = 1,90 \cdot \sqrt{E \cdot f_y} \cdot t^2 \tag{2.28}$$

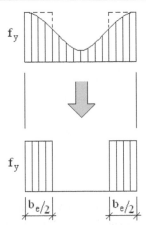

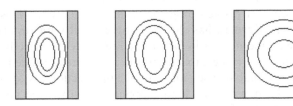

Figure 2.16 A model for the maximum load-carrying capacity as proposed by von Kármán.

Figure 2.17 The effective width is constant for different plate widths according to the von Kármán hypothesis.

We see that the effective width b_e is directly proportional to the root of the Young's modulus E, and to the thickness t. This is obvious, as these two parameters are directly related to the stiffness of the edge zones, and therefore also governs the load-carrying capacity before buckling of the same. However, what is not as obvious, is that the effective width is inversely proportional to the root of the yield stress f_y. But if we consider an increase of the yield stress (i.e. we increase the quality of the plate material), then the effective width must decrease in order to "compensate" for this increase in strength (read: in order to obtain a higher critical buckling stress a reduced width is required).

What is even more surprising is that the effective width is *not* dependent of the original width of the plate (for plates with a panel aspect ratio $a/b \geq 1$). If we increase the width of a plate the maximum load-carrying capacity would remain the same (read: the effective width would remain constant, at least according to the von Kármán hypothesis) (Fig. 2.17).

If we study the plates above, having different widths, we see that as the width increases the wider the buckle also becomes, and therefore also the effective width remains constant. In itself would the critical buckling stress decrease as the width increases (σ_{cr} with the inverse of the width in square, and P_{cr} with the inverse of the width), however, the maximum load-carrying capacity would remain constant (Fig. 2.18).

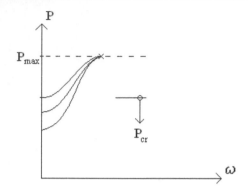

Figure 2.18 For increasing plate widths, the critical buckling load, P_{cr}, decreases, however, the maximum load-carrying capacity, P_{max}, remains constant (according to the von Kármán hypothesis).

There is also a waste of material if one chooses a wider plate than necessary, as the effective width is constant disregarding the width of the plate. This hypothesis that was presented by von Kármán has shown to agree well with results from tests on plates having a large slenderness ratio b/t. In the Eurocode the effective width as a model for the ultimate load-carrying capacity in the post-critical range has also been adopted. The effect of residual stresses and initial imperfections is also taken into consideration (which is influencing the load-carrying capacity, especially for plates having low slenderness ratios). The effective width b_{eff} is calculated as the original width b times a reduction factor ρ (Eq. 2.29):

$$b_{eff} = \rho \cdot b \tag{2.29}$$

where the reduction factor ρ is dependent of the yield stress f_y and the critical buckling stress σ_{cr} (Eq. 2.30):

$$\bar{\lambda}_p = \sqrt{\frac{f_y}{\sigma_{cr}}} \quad \Rightarrow \quad \rho = \frac{(\bar{\lambda}_p - 0,22)}{\bar{\lambda}_p^2} \tag{2.30}$$

There is a certain difference between the effective width of smaller plates calculated according to the von Kármán hypothesis, and the same calculated according to Eurocode, but this difference diminishes as the plates become wider. In general the von Kármán expression overestimates the effective width in relation to the results according to the Eurocode. Table 2.1 compare the results for the *effective width* of different plates, between the von Kármán expression and the Eurocode (under the condition of $f_y = 360$ MPa, $t = 10$ mm, and $k = 4$, i.e. $a/b \geq 1$).

The discussion so far – with respect to the critical buckling stress and the maximum load-carrying capacity in the post-critical range – has been focused on ideal plates without the presence of residual stresses and initial imperfections. We saw, however, already in the discussion above regarding the effective width, that there existed a difference between the von Kármán expression (which neglected these effects) and the Eurocode. As will be seen in the following, these effects have the greatest influence

Table 2.1 A comparison between Eurocode and the von Kármán hypothesis regarding the effective width for different plates.

	Plate width				
	0.5 m	1.0 m	2.0 m	4.0 m	
Eurocode	367	413	436	448	[mm]
von Kármán	459	459	459	459	[mm]

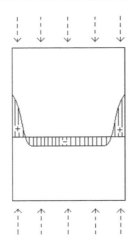

Figure 2.19 Residual stress distribution in a hot-rolled and welded plate.

on the critical buckling stress, and not so much on the ultimate load-carrying capacity (as also was pointed out earlier). The residual stresses in a hot-rolled and afterwards welded plate (e.g. where the edges are fixed to stiffeners) are having the following typical distribution over the plate width (Fig. 2.19).

The compressive residual stresses in the central part will be added to the externally applied load, which will result in a lower critical buckling stress (relative to the classical buckling theory). However, as these residual stresses are self-balancing (read: an inner stress state in equilibrium – the compressive force is equal to the tensile force) will their effect on the maximum load-carrying capacity be minimal (i.e. no major difference between plates having different magnitude of residual stresses). The same applies to the effect of initial imperfections (i.e. deviations from an ideally flat plate) – the critical buckling stress is lowered, but the maximum load-carrying capacity is more or less unaffected. An initial out-of-plane imperfection will gradually increase as the compressive loading is increased, and thus produce a softer transition at the bifurcation point, however, not affecting the ultimate load-carrying capacity as shown in Fig. 2.20.

The sudden out-of-plane buckling at the bifurcation point (according to the classical theory) will not – at least not so dramatically – occur for real plates having initial imperfections and residual stresses coming from production and welding, Instead the buckling will come more soft and gradual. The influence from initial imperfections tends to mask the effect coming from residual stresses, as they have the same effect

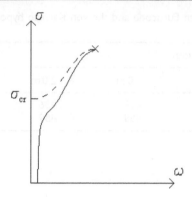

Figure 2.20 The stress/displacement relationship taking initial out-of-plane imperfections into account.

on the behaviour. To separate these effects from each other at a load-testing situation can therefore be difficult. In the codes these effects are taken into consideration as a combined effect, and do not have to be analyzed further by the designer (given that the tolerance levels are held, that is).

Chapter 3

Box-girder bridges

3.1 Introduction

Concerning large thin-walled plates subjected to axial compression, a modern box-girder bridge represents the outermost knowledge by bridge designers of today. For these bridges, high demands are made upon the choice of plate thickness and the position of stiffeners, in order to minimize the possible buckling risk. For the aerodynamically shaped bridge cross-sections that are built today, this knowledge concerning buckling is drawn to the almost extreme perfection.

The box-girder bridge as a successful concept has existed already since the Britannia Bridge was built over the Menai Straits in 1850, but the development up to the modern high technological constructions of today has not been painless, to say the least. Even if the Britannia Bridge marked the introduction of a new and successful concept, the box-girder bridge soon came into the background of the more optimal truss bridges (in terms of the amount of material) – it was first after the Second World War that the box-girder bridge concept once again came into use as a competitive alternative. In the 1960s the development really accelerated, due to the introduction of large rolled plates in combination with the technique of automatic cutting and welding, which did enable for the production of larger bridge cross-sections. The Zoo Bridge, that was built in 1966 in Köln, Germany, had a main span of 259 m – the longest box-girder bridge span in the world at that time – became the start (read: re-start) of this "new" bridge concept for the future, especially as one also had used an unusual erection technique, namely the free cantilevering method. Both the cantilevering erection method as such, and the box-girder bridge as a general idea, came to be questioned very strongly after some bridge collapses during the years 1969–1973 – collapses that we will study more in detail in the following, but first we will look longer back in time to see why the Britannia Bridge was so unique, and then see what can be learnt from the bridge collapses that happened around 1970.

3.2 The Britannia Bridge

In the year 1850 the railway bridge over the Menai Straits in Wales was completed. The bridge was called the Britannia Bridge, and had the longest span in the world for railway traffic. The designer was Robert Stephenson, son of George Stephenson, the famous railway engineer (inventor for example of the record breaking locomotive "The Rocket"). It was the father who in 1838 had suggested that the railway line between

London and Chester should be extended into Wales to the port of Holyhead on the island of Anglesey (to accommodate for the traffic by ferry between England/Wales and Ireland), and so it was decided.

The Britannia Bridge was a continuous box-girder bridge (having two separate tubes) with a total length of 460 m over four spans. The two end spans were 78 m long, and the two central spans were 152 m (often is the length of these main spans said to be 140 m, but that is only the free distance between the towers, not the supported length), see Fig. 3.2.

Figure 3.1 Great Britain and Ireland.

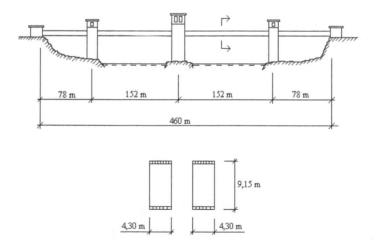

Figure 3.2 The Britannia Bridge – elevation and cross-section.

The somewhat excessive height of the towers was due to the original intention of having the tubes suspended with chains, but this idea was abandoned after some tests (which will be described later in this text) that showed that the tubes were strong enough to carry all load by themselves. (In a suspension bridge having a stiffening girder strong enough to take all loads, the cable becomes superfluous. A flexible cable – that changes form during loading – can only contribute to the load-carrying capacity if the stiffening girder is also flexible. Stephenson was quite rightly so aware of – after the tests – that this was not the case in his suggestion for the bridge.)

The two end spans were assembled supported by falsework, but the two main span tubes were assembled on shore (see Fig. 3.3) close to the bridge site.

To erect the tubes for the main spans, the tidal water was used in an ingenious manner; the tubes were lifted from their position on shore by the help of pontoons, and then shipped out to the bridge location. By the help of hydraulic presses the tubes were then lifted up to the towers – which had been prepared with temporary channels in the exterior face to accommodate for the tubes (which, of course, were longer than the free spacing between the towers), see Fig. 3.4.

The separate tubes were then joined together by lifting one end of a tube up, and then connecting the opposite end with the adjacent tube (see Fig. 3.5). In this way continuity in the system was achieved, which had never been done before for a multi-span girder bridge. By lifting the ends before the joining, and by doing so prestressing the structure, continuity was achieved also for the self-weight, not only for the traffic load (which had been the case if the ends would have been joined without the lifting procedure). Stephenson had designed the box-girders as simply supported, but knew that extra safety and load-carrying capacity would be achieved through this measure, not to forget also a reduced deflection.

Figure 3.3 The two main span tubes were assembled on shore.

Figure 3.4 Preparation for the erection of one of the main span tubes.

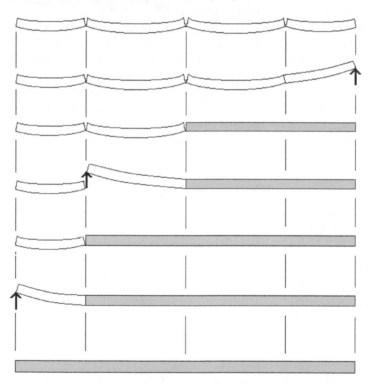

Figure 3.5 By lifting the ends of the tubes before joining them together the tubes became prestressed, i.e. acting as a continuous system.

The moment distribution for self-weight (and additional traffic load) with and or without prestressing shows the great difference in behaviour and stress level (i.e. between a simply supported system and a continuous) in Fig. 3.6.

The cross-section of the box-girders is stiffened in top and bottom with a number of closed cells (small box sections, parallel to the longitudinal axis, integrated with

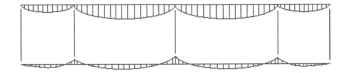

Figure 3.6 The moment distribution before and after prestressing.

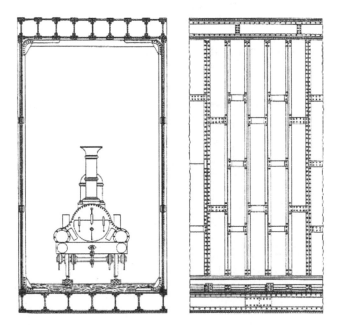

Figure 3.7 Cross-section and side view of the tube.

the large cross-section), eight in the top flange and six in the bottom flange. These cells, which are large enough to accommodate for the re-painting inside, are made out of plane plates, angle irons (i.e. L-profiles) and cover plates, which all have been riveted together in pre-punched holes. These closed cells act as efficient stiffeners with respect to normal stress buckling (due to the bending moment the girders are subjected to) – both for the cross-section as a whole, and also for the small-scale cells (i.e. local buckling of the plane plates in between the angles). One could also see (Fig. 3.7) the cells as flanges for the box-girder cross-section, consisting of double plates, vertically stiffened inside. In the bottom flange the cells also have to carry the transverse (vertical) loading from the railway traffic.

The sidewalls are made up of plane, rolled plates of wrought iron (having a thickness of 16 mm or less), which have been spliced together in the vertical direction by T-profiles, and in the longitudinal direction by cover plates (which had to be bent at the edges in order to wrap over the flanges of the T-profiles). The longitudinal splices are shifted in position in between adjacent plates. In Fig. 3.8 the individual parts are separated from each other in order to show the built-up system.

As the vertical splice is also stiffened inside with a T-profile, the resulting vertical stiffener is made double-sided, and thus very strong (Fig. 3.9).

Even if the compact flanges take the major part of the bending moment, the walls do contribute to some extent, which make them also subjected to normal stresses that can be a local buckling risk. The closely spaced T-profiles (610 mm), together with the chosen wall plate thickness, are, however, a sufficient barrier against normal stress buckling, even if horizontal stiffeners are the normal choice today. In a box-girder bridge, having a cross-section depth of over nine meters, the expected wavelength of a horizontal buckle (in the post-critical range) in the compression zone would be several meters long (approximately 2/3 of the depth, given that the web is unstiffened). This

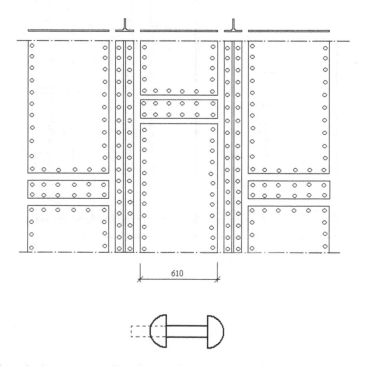

Figure 3.8 The individual plates and T-profiles of the tube wall before being riveted together.

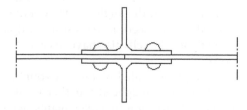

Figure 3.9 The vertical splice T-profiles provided a strong vertical stiffening of the wall.

fact does show that the close spacing between the vertical stiffeners of the Britannia Bridge is more than sufficient. Vertical buckling of the web plate due to concentrated loads is also not a problem, as the load from the trains is introduced in the bottom part of the box, and thus creates only vertical tension as the load is transferred into the web plate. With respect to shear buckling of the walls, the chosen vertical stiffeners are in good agreement with the principles regarding how this type of buckling phenomenon is handled today – crossing diagonals, in tension and compression, in each panel in between vertical stiffeners carries the shear stress flow. The wavelength here is even longer than for normal stress buckling – approximately $1,25 \times h$ – which makes the closely spaced vertical stiffeners also act as a sufficient barrier against shear buckling. However, the choice of vertical stiffeners in the Britannia Bridge was criticized by the Russian bridge and railway engineer D. J. Jourawski. He claimed that if the stiffeners had been positioned in the compression diagonal, having an inclination of 45° towards the longitudinal axis, the efficiency would have been much greater than for the vertical stiffener configuration that was chosen by Stephenson.

When the trains were to pass the bridge, the tubes could obviously not be stiffened inside by cross-framing, instead the torsional stiffness had to be achieved by a horizontal stiffening truss in the roof, and in addition, all four corners of the box were stiffened in order to ensure frame action.

It was the tough requirements from the Admiralty regarding free sailing space in the horizontal and vertical direction that made the choice of a (suspended) girder bridge inevitable (suspended at least according to the original concept). The demands were 32 m free height and 137 m free width in the two sailing channels underneath the bridge. In addition the channel was to be kept open for the passage of ships during the construction. These demands taken together made it impossible to consider the otherwise obvious solution, which would have been a cast iron arch bridge. Besides thinking in terms of a totally new concept for the bridge, they had also to abandon the thought of cast iron as a possible construction material. Cast iron is as strong in compression as (mild) steel, but much more brittle (i.e. less ductile), which makes it suitable for arches where the compression forces are dominating, but less suitable for structures where tension is also present, which definitely would be the case here. Wrought iron, however, is as tough in both compression as in tension, and not as brittle as cast iron. The reason why wrought iron was not as common and competitive as a construction material for bridges was the much higher cost and the fact that it was more difficult to prepare the plates into desired shapes. Cast iron profiles are made directly from the moulding form, which is a huge advantage, however, thus receiving a brittle material. Stephenson pondered on the solution of having two parallel I-girders standing next to each other – clearly inspired by the simple I-girder bridge concept that was so common at the time for short and medium span bridges – but having the top and bottom flange here connected to each other. The shorter I-girder bridge spans were without exception made in cast iron, and had a maximum span length of 15–20 m. The medium span bridges had extra stiffening bars of wrought iron on the tension side, and were approximately 30 meters long. For longer spans the girder bridge had to be suspended with wrought iron eye-bar chains. Stephenson was fully aware of the flexibility of these latter structures (for wind and heavy loading, and not to forget about marching troops!), and knew that his structure had to be a (suspended) stiff girder bridge.

The strength and stiffness of deep plates was confirmed to Stephenson as he was reported about an incident when the steam vessel Prince of Wales had been launched. The ship had accidentally become hanging with the bow in the water and the stern still supported by the slipway on shore, which made the ship's hull becoming simply supported over a length of more than 30 meters, however, without giving any major damages to the hull.

Together with the ship-builder W. Fairbairn and the mathematician E. Hodgkinson, Stephenson carried out a number of experiments in order to determine the action and load-carrying capacity, not only of rectangular cross-section girders, but also of circular and elliptical girder tubes. It was mainly the strong concern he had regarding the wind loading that made him consider the two latter choices (which were more aerodynamically shaped). As the first thought of a possible cross-section shape went to these last mentioned profiles, it is easy to understand where Stephenson did get the idea for hollow tubes, as the trains had to pass inside somehow. Girders in a model scale were tested in three-point bending (i.e. simply supported with a point load in mid-span). They found very soon that the rectangular shape was superior to the other two, mainly due to the buckling tendency of the latter. The flange in a rectangular cross-section is both larger and more efficient than for a circular or an elliptical cross-section. It was also in these tests that they found that the flange in compression had to be shaped with closed cells in order to minimize the risk of buckling, and by doing so ensuring a load-carrying capacity as close to the base material as possible. Fairbairn, that has gone to history as a methodical and scientific researcher concerning the different investigations he did undertake, did comment:

> "Some curious and interesting phenomena presented themselves in the experiments – many of them are anomalous to our preconceived notions of the strength of materials, and totally different to anything yet exhibited in any previous research. It has invariably been observed, that in almost every experiment the tubes gave evidence of weakness in their powers of resistance on the top side, to the forces tending to crush them."

Fairbairn was quite clearly referring to earlier experiments on cast iron girders, which had without exception gone to failure on the tension side (as a brittle fracture), while these tests on wrought iron girders had the failure coming on the compression side (as normal stress buckling). Hodgkinson made the same observation:

> "It appeared evident to me, however, that any conclusion deduced from received principles, with respect to the strength of thin tubes, could only be approximations; for these tubes usually give way by the top or compressed side becoming wrinkled, and unable to offer resistance, long before the parts subjected to tension are strained to the utmost they would bear."

Hodgkinson knew that the simple mathematical expressions which traditionally had been used – where the maximum load-carrying capacity was derived from the ultimate tensile strength of the material – definitely overestimated the capacity. It was also during this work, together with additional buckling tests they made on plane plates, that Hodgkinson found that the maximum load-carrying capacity, with respect on buckling of an axially loaded plate, was directly proportional to the thickness of the plate in cube. That this is the case is easily derived if one transforms the expression for the critical buckling stress into an equivalent axial load (see Eq. 3.1):

$$\sigma_{cr} = k \cdot \frac{\pi^2 \cdot E}{12 \cdot (1 - v^2) \cdot \left(\frac{b}{t}\right)^2}$$

$$P_{cr} = \sigma_{cr} \cdot b \cdot t \Rightarrow \underline{\underline{P_{cr} \sim t^3}} \tag{3.1}$$

Hodgkinson had clearly an early and genuine understanding of the ultimate capacity of axially loaded plates, even though that he stated that the capacity was limited by buckling (and therefore excluded the post-critical reserve strength). And by excluding the reserve effects in the design of the Britannia Bridge, the structure came to have an extra safety level that to some extent explains why it was possible to raise the traffic load without having to strengthen the bridge.

Stephenson and Fairbairn also found in the experiments, that the clamping force in between the connected plates was sufficiently high, that the friction was not exceeded due to shearing in the joints. The deflection one could expect from the bending deformations of the box-girder spans was consequently only coming from elastic deformations of the material, and not by additional movements from shearing making the rivets coming in contact with the rivet hole edges. This fact was also confirmed as the deflection was checked during the passage of trains. Despite small expected deflections one did still choose to provide the centre spans with a camber of 23 centimeters.

The concept of having a closed box-girder section, where especially the walls were made out of solid plates, made the bridge become extremely expensive due to the large amount of material that was used. The labour cost in the 19th century was low, but the cost of the material was high. Instead the development went towards the more lighter and material-saving truss structures – a development that started in the USA. Smaller truss bridge spans in wood became increasingly longer as more and more parts were replaced by iron. A special concept that was invented by Pratt, where the diagonals were positioned so that they always became in tension during loading, was very competitive, as material was saved due to the non-existent risk of global buckling of these diagonals. Stephenson, however, carried on by designing and developing his heavy and robust box-girder bridge concept with great success during the 1850s. Among other bridges, he designed for example the Victoria Bridge (1859) over the St Lawrence River near Montreal in Canada. This bridge was the longest bridge in the world at that time (1.8 kilometers long), and did have a maximum span length of 100 meters. The bridge had to be rebuilt after some time though, as the smoke from the locomotive did trouble the passengers – the walls had to be opened up!

Even though that the Britannia Bridge is said to be the start of the modern era concerning box-girder bridges, the fact is that a box-girder bridge was built in Conway – 25 kilometers east of Menai Straits, also on the Chester-Holyhead railway line – already a year earlier. This bridge, with a single span of quite impressive 122 meter, served as a prototype for the Britannia Bridge – here they had the opportunity to test the design and erection method. It was from the Conway Bridge that the idea came for the camber of the Britannia Bridge (the Bridge at Conway was made completely straight, which made

Figure 3.10 The box-girder bridge at Conway, 25 kilometers east of the Menai Straits.

the deflections from self-weight and traffic become visible). It was also some important experiences made from the lifting process, even though that the lifting height at the Conway Bridge was only 5.5 meter, and not 36 meter as was the case for the Britannia Bridge (Fig. 3.10).

In 1899, the Conway Bridge was reinforced with two extra inner supports, approximately 15 meters from each abutment, which made this bridge also continuous. The extra supports can dimly be seen in the photo above, in between the castle tower supports.

The Britannia Bridge was the longest box-girder bridge in the world (given the maximum span) up until the period after the Second World War, when several new and modern box-girder bridges were built, mainly in post-war Germany. As has been mentioned before, the Britannia Bridge did carry heavy traffic loading without having to be strengthened (even though the traffic load increased gradually), and continued to do so until 1970, when the wooden cladding of the roof accidentally was put on fire. After 120 years of faithful service, the Britannia Bridge had to be completely rebuilt, and the solution was to transform the structure into a truss arch bridge (Fig. 3.11) (an impossible solution as it was in 1850!). The bridge is carrying today, besides railway traffic, also highway traffic on the roof of the original bridge.

The original bridge was truly a record-breaking structure, having many characteristics that made it identical to a modern box-girder bridge:

- large spans,
- large dimensions in general (but perhaps more a matter of large depth than width),
- longitudinally and transversely stiffened plate panels,
- prestressing of the box-girders,
- a continuous system,
- a spectacular erection,
- the use of camber,

Figure 3.11 The Britannia Bridge of today (after being redesigned due to the fire in the early 1970s).

– the use of a new material,
– many competent and skilled persons involved.

P.S. Even if the Britannia Bridge no longer exists in its original shape, the bridge at Conway is still standing intact!

3.3 Collapses

In the four-year period in between 1969 and 73 there were not less than five box-girder bridge collapses during construction:

– The Fourth Danube Bridge in Wien
– The Cleddau Bridge in Milford Haven
– The West Gate Bridge in Melbourne
– The Rhine Bridge in Koblenz
– The Zeulenroda Bridge in former East Germany

Four of the bridges were situated in Europe (Fig. 3.12); while one bridge was situated outside Melbourne in Australia (Fig. 3.13).

Prior to these collapses, there was first the Britannia Bridge which showed that the concept of box-girder bridges successfully had been in use already since 1850, and in addition there was the fact that modern box-girder bridges had been built during the 1950s and the 1960s, and without any major incidents occurring. The theory for plate buckling was since long well-known, however, complete design recommendations for box-girder bridges were missing in the codes, which can explain why a certain practice by usage was adopted, especially concerning the design of stiffeners. We will in the following sections also see that the chosen erection methods did induce stresses in the bridges that they to the highest certainty were not designed for.

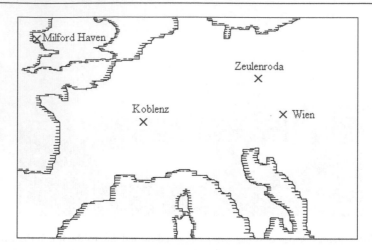

Figure 3.12 Central Europe.

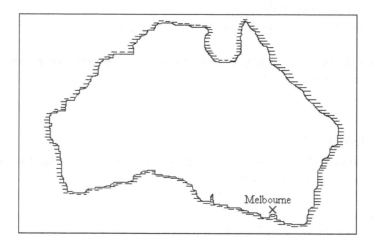

Figure 3.13 Australia.

3.3.1 *The Fourth Danube Bridge*

On the evening of the 6th of November 1969, there was some unexpected buckling occurring during the final stages of construction of the Fourth Danube Bridge in Wien, Austria. Far away from the actual bridge site people could hear, with an interval of five seconds, three loud bangs, almost like explosions, and they all came from the bridge that had buckled under excessive loading. As the bridge not yet had been taken into service, the loading could only be coming from self-weight, which made it quite mysterious.

The Danube Bridge was a continuous box-girder bridge in three spans, having haunches over the inner supports. The total bridge length was 412 meters, with a main span of 210 meters (Fig. 3.14).

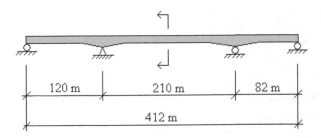

Figure 3.14 Elevation of the Fourth Danube Bridge.

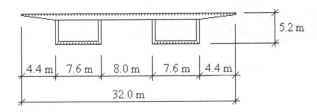

Figure 3.15 Cross-section of the Fourth Danube Bridge.

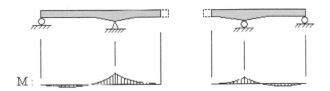

Figure 3.16 Moment distribution during the erection process of the bridge, using the free cantilevering technique for the mid-span part.

The bridge was a twin-box girder bridge, having a 5.2 meter deep and 32.0 meter wide cross-section in mid-span. The top and bottom flanges were stiffened in the longitudinal direction with simple flat plates, see Fig. 3.15.

In order to minimize the disturbance for the boat traffic on the river Danube, one had chosen the free cantilevering erection method. By successively assemble the box sections to the ends of the cantilevers – instead of using supporting falsework – there is a high demand for proper anchorage at the end supports, as well as a sufficient load-carrying capacity of the cantilevers. As the length of the cantilevers are increasing, the moment and shear over the inner supports also increase – stresses that exceed what is later the case for the final bridge (Figs. 3.16, 3.17) (compare also the drawing to the photo of the Danube Bridge during construction).

When the two cantilever ends met in the middle, the final section had to be adjusted due to the temperature deformations the bridge had been subjected to during the day. The weather had been sunny and warm, which had made the bridge, in addition to

Figure 3.17 Only a short gap remaining before the final closing of the two cantilevers.

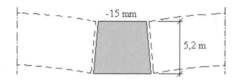

Figure 3.18 The final closing section in the centre part had to be adjusted due to temperature deformations in the bridge.

the elastic bending deformations, bend down a little bit extra. The final closing section (Fig. 3.18) had to be shortened 15 mm at the top (due to this additional temperature deformation) in order to fit into the gap between the cantilever ends.

Towards the evening, when the temperature dropped, the temperature deformations reversed, however, as the final section already had been installed, a constraint was introduced. This constraint did induce a tensile force in the top flange (that was prevented from shortening), and, as a consequence, compressive stresses was introduced in the lower flange. One could also say that the cantilever ends were prevented from rotating back to their original position. In addition to this effect, there were the already high compressive stresses introduced in the lower flange (that we will concentrate on in the following) because of the chosen erection method. Two gigantic cantilevers had been joined together; however, the moment distribution was not yet the one that could be expected from a continuous system (Fig. 3.19). And in order to achieve this, one had to adjust the inner supports down, so that the undesired moment distribution was levelled out.

In the late afternoon, when the two cantilever ends had been joined, there was, however, not enough time left to lower the inner supports, instead it was decided to wait until the next working day. Due to the temperature deformations described above, there was a situation in the evening that had given an increase in the temporary

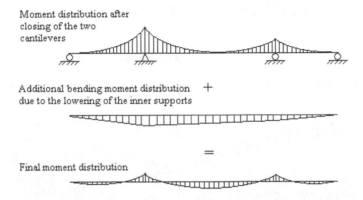

Figure 3.19 The moment distribution that would be the result following the intended lowering of the two inner supports.

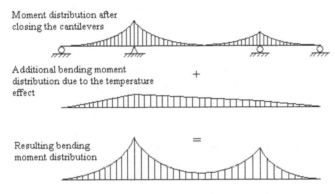

Figure 3.20 The resulting moment distribution due to the temperature effect (before the intended lowering of the inner supports).

moment distribution, instead of the desired decrease that would be the case if one had lowered the supports (Fig. 3.20).

By the unfortunate decision of not lowering the inner supports immediately, and the unlucky effect from the temperature drop, there was a situation – the night of the 6th of November – where the lower flange was subjected to compression over the entire length of the bridge (due to the negative bending moment). This fact, together with the less good decision of choosing flat bar stiffeners, was the reason for the buckling of the bridge. A flat bar stiffener is very easily deformed during welding (due to the welding deformations coming from the excessive heat input), and can also be deformed due to more or less unavoidable hits and damages during assembly. A deformed flat bar stiffener has a markedly reduced global buckling strength capacity (see further the discussion of the Zeulenroda Bridge, section 3.3.5). As the lower flange of the Danube Bridge was subjected to excessively high compressive stresses, especially in the zones where the final bending moment after lowering of the supports would have been low (compare Fig. 3.19) – i.e. near the zero-moment positions – it was not surprising that

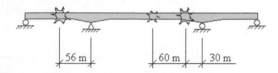

Figure 3.21 Buckling did occur in three locations due to the excessive bending moments.

Figure 3.22 The parts having the major buckles – close to the inner support of the main span, and in the centre part of the first span.

buckling did occur there (the buckle farthest to the right that is marked in Fig. 3.21). The lower flange stiffeners at these zero-moment positions are not designed for any larger stress levels (at least not for normal in-service loadings). The other two buckles did occur in zones where the final (desired) moment distribution would be positive, i.e. in areas where the lower flange would be subjected to tension stresses (read: in mid-span regions). However, as the moment distribution in these areas was reversed, the lower flange was subjected to compression (instead of tension), and thus the stiffeners buckled due to excessive (compressive) loading (Fig. 3.21).

The buckle in the central region of the inner span came where the final section was located, however, this piece was not severely damaged (even though that this section was not as strong as the rest of the bridge, due to the fact that it was assembled without the lower and upper flange plates, i.e. only having the four vertical web plates). The other two buckles – the one in the main span close to the inner support, and the one in the centre part of the first span – were, however, so large that more or less true hinges were formed in these positions (inwards going buckles in the vertical web plates, and outwards going – V-shaped – buckles in the lower flanges). In the detail Fig. 3.22, one could see the proportion of the damages in these two locations.

In the photo of the elevation (view) of the bridge (see Fig. 3.23), the deformed and upwards bent span to the left can be seen (compare the deformation with the mechanism in Fig. 3.26).

As the continuous girder system was two times statically indeterminate, the bridge was able to withstand the formation of the two hinges. The girder system now became statically determinate, with the right part (1) holding up the centre part (2), which in its turn was holding up the left part (3) (Fig. 3.24).

Through this release of the constraint due to the statically indeterminacy, the girder system now received the moment distribution (in principle) that was the intended after lowering of the supports (Fig. 3.25).

Figure 3.23 The downward deflected main span and the upward bent first span (to the left).

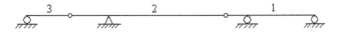

Figure 3.24 The statically determinate system after the buckling had occurred.

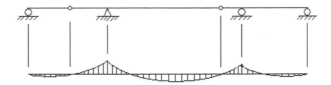

Figure 3.25 Bending moment distribution after that the buckling had occurred (in principle similar to the intended after the planned lowering of the two inner supports).

Figure 3.26 The mechanism (read: collapse) of the bridge if the closing section also had been severly damaged.

However, if in addition the closing section also had been damaged to the extent of the other two buckled parts – so that a hinge also had been produced there – then the collapse (read: a mechanism) had been a fact (Fig. 3.26).

Thanks to the inbuilt damage tolerance of the system, the bridge could be saved, even though it was heavily distorted, and was later taken into use again (after replacement of the damaged parts). The rebuilt bridge, which today goes under the name of the Prater Bridge, obtained a permanent settlement of about 700 mm in mid-span of the centre span (read: a reduced camber) as a direct consequence of the damages.

3.3.2 *The Cleddau Bridge*

Less than seven months after the buckling incident of the Danube Bridge, the Cleddau Bridge, near the seaport of Milford Haven, Wales, collapsed on the 2nd of June 1970 during construction. It is a little like irony of fate that just ten days after the destruction of the Britannia Bridge (that happened the 23rd of May the same year), the Cleddau Bridge collapses – a bridge that would be one of the longest in Europe (having a

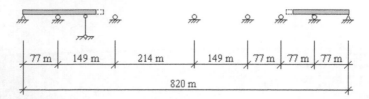

Figure 3.27 Elevation of the Cleddau Bridge during erection (prior to the collapse of the right-hand part, on the south side).

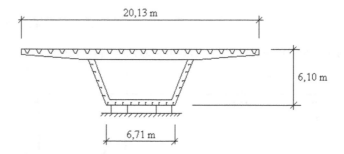

Figure 3.28 Cross-section of the Cleddau Bridge.

maximum span of 214 meters). The predecessor and the inheritor, are made unfit for use almost at the same time, and top of all, they were both located in Wales.

The Cleddau Bridge was to be a continuous box-girder bridge in seven spans, with a total bridge length of 820 meters (Fig. 3.27) (in the re-designed bridge – after the collapse – the centre part was rebuilt though into a cantilever bridge, having a reduced span length as well).

After the end spans on either side had been erected (using false-work), the free cantilevering erection method was used for the continued assembly. For the north end (to the left in the elevation above) it was decided to use a temporary support in mid-span of the second span, while on the south side (to the right in the elevation) it was decided to manage without, due to the shorter span length. The trapezoidally shaped cross-section of the bridge had a lower flange width of 6.71 meter, an upper flange width of 20.13 meter, and a depth of 6.10 meter (Fig. 3.28).

The designer, Freeman, Fox & Partners, had with great success a couple of years earlier, used a trapezoidally shaped cross-section at the construction of the Severn Bridge (a 988 meter long suspension bridge) and at the construction of the Wye Bridge (a 235 meter long cable-stayed bridge), and wanted to continue with this successful concept. The box-girder bridge cross-section was here, in contrast to the Danube Bridge, stiffened in the longitudinal direction by the use of more buckling stiff and torsionally rigid stiffeners. The assembly of the second span on the south side of the bridge, had at the time of the collapse cantilevered out 61 meters, and the final section (Figs. 3.29, 3.30) (closing the gap of 77 meters) was just to be launched out, as the bridge buckled over the inner support, and the huge arm fell 30 meters to the ground (killing four people).

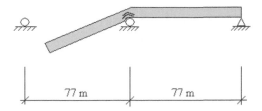

Figure 3.29 Collapse of the second span on the south side.

Figure 3.30 The cantilever arm of the second span buckled over the inner support and fell to the ground.

In contrast to the Danube Bridge – which luckily did receive its damages first after the system had become statically indeterminate – the Cleddau Bridge had no such extra inbuilt safety during construction, as the system was statically determinate. As soon as this hinge was produced (read: the buckling at the inner support) a mechanism became the result. At the investigation it was established that an inadequately stiffened diaphragm had initiated the buckling over the support.

Diaphragms are made up of a solid plate in box-girder bridges (having a centrally located man-hole though, for the passage of personnel during inspection), and these

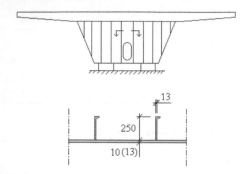

Figure 3.31 The stiffened diaphragm plate of the box-girder over the inner support.

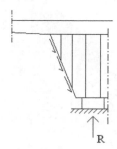

Figure 3.32 The shear load from the inclined webs is transferred down to the bearings.

are positioned at a regular interval in the longitudinal direction (say 8–12 m) in order to ensure a sufficient torsional stiffness of the cross-section. The diaphragms are stiffened, especially over the supports, as they are in these positions subjected to large concentrated forces. In the Cleddau Bridge, the diaphragm plate was stiffened with so-called bulb flat stiffeners, having the dimension of $250 \times 13 \, \text{mm}^2$. The plate itself was 10 mm thick in between the bearings, and 13 mm locally over the bearings and to the inclined webs (Fig. 3.31).

A diaphragm over a support has many functions:

– Transfer the shear load, that is coming from the inclined webs, down to the bearings,
– Carry the reaction forces,
– Act as a vertical stiffener of the web plate,
– Act as a stiffening girder during the replacing of a bearing,
– Transfer the horizontal forces from wind and traffic down to the bearings.

The first function mentioned above, means that the diaphragm has to act a stiffening (transverse) girder that is loaded at the edges with the inclined shear force (Fig. 3.32).

As a comparison – in order to better understand the action of a diaphragm having inclined edges – we study a diaphragm for a rectangularly shaped cross-section (Fig. 3.33).

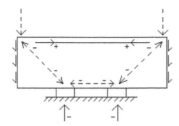

Figure 3.33 As a simplified model showing the load distribution – a rectangular diaphragm plate.

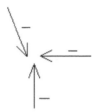

Figure 3.34 Equilibrium due to the inclined diagonal force is met by a horizontal balancing (compressive) force.

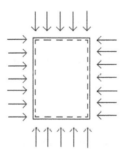

Figure 3.35 Locally, close to the bearings, the diaphragm plate is subjected to two-axial compression.

The shear load must also here be transferred from the edges (where the web plates are connected) to the bearings. In a simplified manner we assume that all loads are introduced in the upper part of the diaphragm, and are then transferred down to the bearings through a diagonal, that is in compression. Even if the diaphragms in general are large, there is a local concentration of forces in compression over the bearings. Besides the vertical reaction force and the diagonal force, the inclined diagonal force has also to be balanced by a horizontal force in this region (Fig. 3.34).

The diaphragm plate is thus subjected to a two-axial compression locally over the bearings (Fig. 3.35).

The critical buckling stress for a plate loaded in two-axial compression, follows a linear relationship according to (Eq. 3.2):

$$\frac{\sigma_x}{\sigma_{x,cr}} + \frac{\sigma_y}{\sigma_{y,cr}} = 1,0 \quad \Rightarrow \quad \sigma_x = \sigma_{x,cr} \cdot \left(1 - \frac{\sigma_y}{\sigma_{y,cr}}\right) \tag{3.2}$$

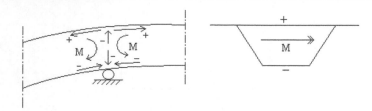

Figure 3.36 Additional vertical reaction force (in the diaphragm plate) was induced due to the curvature effect.

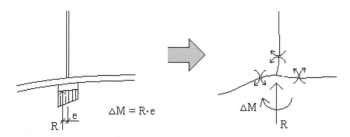

Figure 3.37 Eccentric loading of the bearing pad induced local bending of the lower flange plate and of the diaphragm plate.

The relationship could also be expressed through the buckling coefficient k (Eq. 3.3):

$$k_x = 4.0 \cdot \left(1 - \frac{\sigma_y}{\sigma_x}\right) \tag{3.3}$$

As the loading in the transverse direction is increasing, the critical buckling stress – for the higher loading in the x-direction – will decrease (i.e. the buckling coefficient will become less than 4.0).

Besides the two-axial stress state – due to the transfer of the shear force from the inclined webs – there were a number of other factors that even more increased the stresses over the support. There was, for example, the additional compression in the diaphragm plate coming from the curvature effect (Fig. 3.36).

In addition did the curvature induce a local eccentricity, as the 40 mm wide bearing plate became eccentrically loaded. This eccentricity made the lower flange (of the box-girder) and the diaphragm wall become subjected to a bending rotation (Fig. 3.37)

In the investigation it was also found that the "holding-down bolts", which at a later stage of the erection (cantilevering of the third span) should prevent lifting of the bearings, had been tightened. The effect of this tightening was most certainly that the support became fixed, which additionally increased the vertical reaction force over the supports (Fig. 3.38).

All of these effects taken together did create an excessive loading of the diaphragm plate locally over the supports, and this made the plate and the stiffeners buckle, followed by the buckling of the web plates, and then the collapse of the entire cross-section. The code at that time was quite clearly inadequate with respect to proper design

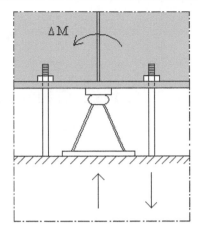

Figure 3.38 Fixing the bolts (which were needed in holding down the box-girder bridge when cantilevering out the third span) at this early stage meant that an additional reaction force was introduced (due to the clamped condition).

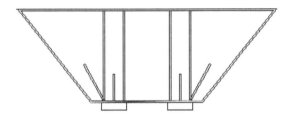

Figure 3.39 The configuration of a modern box-girder bridge diaphragm plate, having extra stiffeners in the highly stressed regions.

of diaphragm plates, and also were the designers not fully aware of the complexity regarding the action of the same. Today the diaphragms are not only having larger thickness and stronger stiffeners, they are also provided with extra stiffeners locally where the stress concentrations are the highest (Fig. 3.39).

3.3.3 *The West Gate Bridge*

The West Gate Bridge in Melbourne was to become the without comparison largest bridge that Australia have ever seen before – four lanes in each direction, and with a total length of almost 2.6 kilometers. The bridge consisted of two parts; approach spans in reinforced concrete of 67 meters each, and a centre part in steel having a total length of 848 meter. This steel section was a continuous box-girder bridge in five spans, having the three centre spans, over the Lower Yarra River, suspended by stay-cables. The main span of 336 meters would be one of the longest in the world for a cable-stayed bridge (Fig. 3.40).

The trapezoidally shaped cross-section had two inclined and two vertical webs, all together forming a three-cell box-girder bridge. Besides contributing to the load-carrying capacity, the vertical webs also functioned as inner supports for the crossbeams in the upper and lower flange (Fig. 3.41).

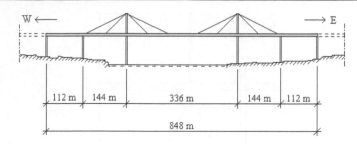

Figure 3.40 Elevation of the West Gate Bridge.

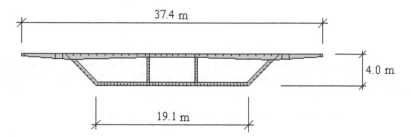

Figure 3.41 Cross-section of the West Gate Bridge.

The choice of designer came once again to be the renowned English consulting engineer Freeman, Fox & Partners, as the competence was judged to be lacking in Australia for such a large bridge. FF&P had also an impeccable and good reputation going back to 1932 being the designers of the Sydney Harbour Bridge.

The construction of the bridge started in April 1968, and there was a hope that the work should be finished to December 1970, however, due to strikes and other delays, the work came to be seven months behind time schedule already by the end of 1969. In the beginning of 1970 the original steel contractor was replaced, but the fact was still remaining that construction was much overdue. The stress was definitely not lessening as the message came in June 1970 that the Cleddau Bridge in Milford Haven had collapsed, and that FF&P was the designer also there. FF&P made the best of the situation, trying to explain to the authorities and working personnel that it was a once in a lifetime incident, and that the bridge in Milford Haven was built using the free cantilevering technique, which was not the erection method used for the West Gate Bridge. The first spans on the east and west sides (those having a span length of 112 meter) were just to be erected, and the technique was to build the spans on the ground, before lifting them into position resting on their inner supports. The Milford Haven collapse, however, led to the strengthening of these cross-sections.

The time pressure made the contractor choose a somewhat unusual erection sequence though. In order to save time and reduce the weight for the lifting process, it was decided to build the bridge in two separate halves. One half at the time was then lifted up by the help of hydraulic jacks, and launched horizontally on a sliding girder into position (Fig. 3.42).

The photo in Fig. 3.43 shows the two bridge halves before being joined together.

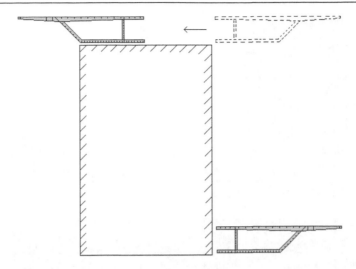

Figure 3.42 The lifting procedure of the two separate halves.

Figure 3.43 The two separated halves, on top of the supports, before being joined together.

Already at the assembly of the first half of the bridge on the ground on the east side, problems had arisen. The free flange edge of the inner part had buckled, as the span lay simply supported before the lifting process, having the entire length free and unsupported. The box-girder half became already on the ground subjected to maximum moment due to self-weight – the lower flange in tension, and the upper flange in compression, which did buckle due to the stresses it became subjected to (Fig. 3.44).

Figure 3.44 The upper (and outstanding free) flange buckled already when the bridge half lay on the ground, supported at the ends (before being lifted).

Strangely enough did they not attend to the problem when the bridge half was still on the ground, instead it was decided to continue with the lifting process, and thereafter deal with the buckled flange edge up in the air. The logical thing had of course been to remove the stresses by putting the bridge back to its original supports, along the entire length of the girder, and thereby unload and straighten the upper flange. And by doing so, they would have been given themselves the chance of also strengthening the flange edge.

The decision to assemble the bridge in two separate halves was the reason why the flange became free and unsupported, which also made it buckle. We will in the following study this free flange edge more in detail, and compare the critical buckling stress of the same, in comparison to the case when the two flanges of the separate halves have been joined together. The protruding free flange was reinforced in the longitudinal direction by bulb flat stiffeners (just as for the Cleddau Bridge in Milford Haven) each 1060 millimeters. At the free edge of the flange, this distance came to be halved. The flange plate and the longitudinal stiffeners were in their turn stiffened by a crossbeam every 3.2-meter in the transverse direction – the cross-beam was also a bulb flat stiffener (460 millimeter deep) (Fig. 3.45).

Let us now study the part of the stiffened flange that is closest to the free edge, having the remaining three edges supported by a longitudinal stiffener and two crossbeams (Fig. 3.46).

The critical buckling stress (for this plate supported on three edges), with respect to an evenly distributed compressive load, will be (Eq. 3.4):

$$\sigma_{cr} = 0.425 \cdot \frac{\pi^2 \cdot E}{12 \cdot (1 - \upsilon^2) \cdot \left(\frac{530}{9.5}\right)^2} \tag{3.4}$$

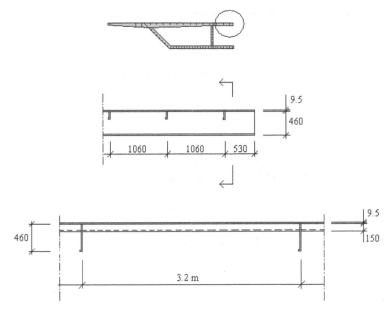

Figure 3.45 Cross-section of the bridge half, with a close-up of the upper flange end together with a side-view of the same.

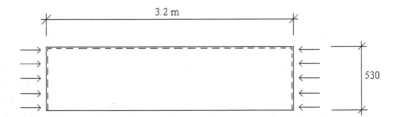

Figure 3.46 The free flange plate in between longitudinal and transverse stiffeners.

As the free edge will be joined with the opposite flange plate (from the other bridge half), the critical buckling stress will increase (i.e. beneficial), even though the width has been doubled (Eq. 3.5 and Fig. 3.47):

$$\sigma_{cr} = 4.0 \cdot \frac{\pi^2 \cdot E}{12 \cdot (1 - v^2) \cdot \left(\frac{2 \cdot 530}{9.5}\right)^2} \tag{3.5}$$

If we now compare the critical buckling stresses with each other, we find that the latter – the one with the doubled width – is 2.35 times larger than the former (Eq. 3.6):

$$\frac{\left(4,0/2^2\right)}{0.425} = \underline{\underline{2.35}} \tag{3.6}$$

Figure 3.47 The width of the flange plate – in between the longitudinal stiffeners – will be doubled, as the two bridge halves are joined together, but still the critical buckling stress will increase.

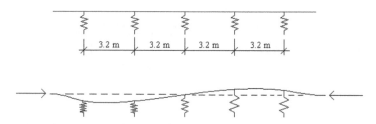

Figure 3.48 Buckling of the longitudinal stiffener, being supported by flexible transverse stiffeners.

The critical buckling stress became more than 50% reduced due to this temporary free edge of the flange, which also was realized already on the ground. However, as has been mentioned earlier, they waited until the bridge was lifted into position before taking care of the buckles, even though the buckles were up to 380 millimeters in amplitude! The reason behind the large amplitude was the fact that also the longitudinal stiffener (closest to the free edge) had buckled, due to the flexibility of the transverse cross-beams (which did not act as rigid supports, as they were flexible cantilevers) (Fig. 3.48).

This flexibility of the transverse cross-beams, made the longitudinal stiffener have a buckling length longer than the distance between the cross-beams. The deflections of the cross-beams were 50–75 millimeter, and it was this flexibility that contributed to the buckling of the plate edge, and also explains why the deflection of the same became as large as 380 millimeter. In addition, a weak splice of the longitudinal stiffener also did contribute to the reduced buckling strength. Each 16th meter the longitudinal stiffeners were spliced, by using a simple single-lap joint, having a rectangular plate, 100×12.5 mm^2, overlapping the two bulb flat stiffener edges. The designer had by this choice of configuration inserted an local weakness of the stiffeners – the overlapping plate was not only smaller in dimension (than the bulb flat stiffeners), it was also eccentrically positioned, and last, but not least, not welded to the upper flange plate, which made it less strong in the transverse direction. This detail had thus a markedly reduced strength of transferring axial load from one end of a longitudinal stiffener to the other (Fig. 3.49).

Being up in the air, there was no chance of unloading the bridge half, instead they had to solve the problem with the buckled plate edge in a different manner. They

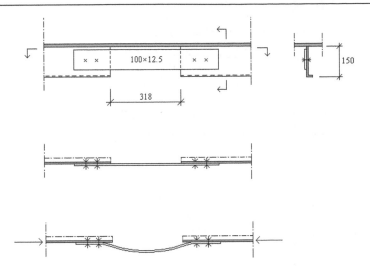

Figure 3.49 The rather weak detailing of the longitudinal stiffener splice (each 16th meter).

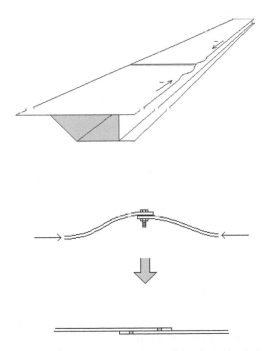

Figure 3.50 By removing the bolts in transverse splices (of the deck plate) the buckles were removed due to the plates slipping under each other.

did then take a bold decision, to quite simply open up transverse splices in order to "relieve" the inbuilt stresses, which were the cause of the buckling – the plates would then slip under each other, and by doing so straighten out the buckled plate. This was a step that worked quite satisfactory for this east span (Fig. 3.50).

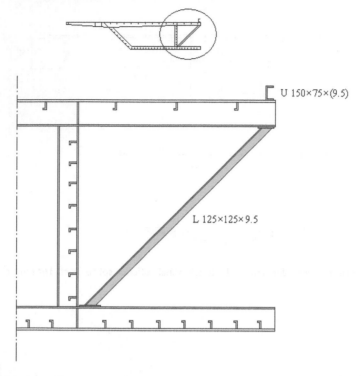

U 150×75×(9.5)

L 125×125×9.5

Figure 3.51 The upper flange edge of the bridge halves on the west side were stiffened by an extra longitudinal, and the transverse cross-beams were supported by an additional diagonal.

New holes had to be drilled (or the enlargement or re-drilling of already the existing ones) in order for the bolts to be inserted again (the slip between the plates is heavily exaggerated in the drawing above, see Fig. 3.50).

For the continued work of the assembly of the bridge halves on the west side, the free flange edge was stiffened with an extra longitudinal, and each cross-beam was supported by a diagonal – all due to the lesson made from the buckling of the east bridge (Fig. 3.51).

At the assembly of the west girder halves in the air, this extra reinforcement did show to be efficient, and there was initially no problem with buckling of the upper flange. Instead another problem arose when the two halves were to be joined together – due to tolerances that were exceeded, and a difference in deflection, there was a vertical gap of 115 millimeter in mid-span between the two halves. There had been a similar problem also for the east span, however, due to smaller distance the discrepancy was evened out by the help of jacks. A distance of 115 millimeter is too much for jacks to even out, so they had to come up with another idea for this west span. It was then that they came up with the in a sense "logical" idea, but alas, so fateful. They took seven concrete blocks of eight tons each, and placed them in mid-span on the half they wanted to deflect down (Fig. 3.52).

The extra strengthening of the free edge – that showed to be sufficient for loading coming from self-weight alone – did show insufficient for the additional loading coming

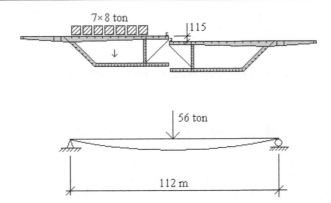

Figure 3.52 Extra loading were put on one of the bridge halves to even out the difference in deflection.

Figure 3.53 The difference in deflection was evened out by the extra concrete blocks (to be seen on the left), but the free edge (of the extra loaded half) buckled, still being stiffened by the extra longitudinal at the free upper flange end.

from this temporary loading of the concrete blocks (the bending moment in mid-span increased some 15–20% relative to the bending moment of the self-weight alone). The difference in deflection evened out, however, the entire upper flange plate buckled (including the longitudinal stiffeners). In Fig. 3.53, the concrete blocks can be seen on the left, as well as the extra longitudinal stiffener at the free edge.

The continued work of the assembly was taken to a halt completely, and for more than a month (!) they discussed about what to do about the situation. Finally it was

Figure 3.54 The "man-made" produced hinge made the bridge collapse to the ground.

decided to go ahead with the method used for the east span, i.e. to open up a transverse splice by removing the bolts. They started with the splice in mid-span and removed the bolts one by one – the difference being, in comparison to the east span, that the buckle now was much larger and the loading heavier. As more and more bolts were removed, the stress increased on the remaining active part of the upper flange of the bridge – not only due to the loss of cross-sectional area, but also due to the gradual lowering of the neutral axis. After 16 bolts had been removed, there was a visible reduction of the buckles (through slipping of the plates), but there was also the effect that some of the remaining bolts were squeezed tight (which made the continued loosening more troublesome). When 37 bolts had been removed the inevitable collapse of the cross-section started – a buckle in the upper flange plate spread in the transverse direction (across the width of the bridge) due to the overloading of the net section, and also the vertical web did buckle in its upper part (where it was subjected to compression). The remaining bolts of the upper flange were cut off due to the excessive shearing forces between the plates, and this was followed by a slow sinking of the left-hand bridge half (compare Fig. 3.53), which now was carried by the right-hand bridge half (the intact half) alone – there had been some connecting of the two halves earlier. However, as the loading of this right-hand bridge half also became too much, it gave after for the excessive loading and the entire bridge fell some 50 meters to the ground – 36 people were killed on this tragic accident the 15th of October 1970. Just as was the case for the Cleddau Bridge in Milford Haven – where the buckling of the diaphragm produced a hinge, that transformed the statically determinate system into a mechanism – so did the "man-made" hinge here in the West Gate Bridge (Fig. 3.54) make the simply supported girder span collapse to the ground (the clearly visible hinge – as folding of the upper flange – can also be seen on the photo in Fig. 3.55).

Every 16th meter one could see a local "ridge" in the upper flange, and this was the result of the weak detail described earlier of the splicing of the longitudinal stiffeners (Fig. 3.56) (can also be seen at the lower end of the photo in Fig. 3.55).

This local damage had nothing to do with the actual collapse of the entire bridge, instead it can be seen as a last and final "death-rattle".

The collapse was caused by a series of faulty decision, which all started by the contractor choosing the rather unusual method of assembling the bridge in two separate halves. Stephenson did show, already back in 1850, that it is not impossible to perform a lift of a 1600-ton heavy bridge section (which was the weight of one of the tubes for the centre span of the Britannia Bridge). If they had decided, for the West Gate Bridge, to carry out the lifting of the bridge as a complete unit, the weight would "only" have been 1200 tons, and then all the problems concerning buckling and difference in vertical deflection would have been gone.

Figure 3.55 The collapsed bridge span on the ground.

At the investigation of the reason for the collapse, it was found, besides the blame of the contractor, that Freeman, Fox & Partners had to take the major part of the responsibility:

- They had not checked the load-carrying capacity thoroughly enough for the assembly (and lifting) of the bridge in two halves.
- Questions regarding the structure as a whole had FF&P been neglecting to answer during construction.
- The inspection on site had been scarce, and carried out by a 23 year old and newly become civil engineer.
- Last, but not least, the main responsibility rested on FF&P for the decision to remove the bolts.

Figure 3.56 The local collapse of the weak splicing of the longitudinal stiffeners (compare Fig. 3.49).

If you ever visit The Monash University, Department of Civil Engineering, in Melbourne, then take the opportunity to study some of the remaining parts from The West Gate Bridge. Some "scrap pieces" have been saved there, outside the building, as a reminder of the importance of remembering bridge failures, and to learn from the mistakes made. On the left in the photo one can, for example, see the buckled splice between the longitudinal stiffeners (Fig. 3.57).

Figure 3.57 Some removed and saved parts from the collapsed West Gate Bridge at Monash University in Melbourne, Australia.

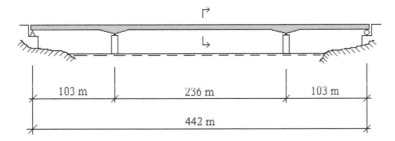

Figure 3.58 Elevation of the Rhine Bridge, near Koblenz.

3.3.4 *The Rhine Bridge*

On the 10th of November 1971, little more than a year after the collapse of the West Gate Bridge, a box-girder bridge over the Rhine near Koblenz, close to the inlet of the tributary river Mosel, fell down during construction, killing 13 workers. The bridge (also called "The Südbrücke") was a continuous and haunched box-girder bridge in three spans, and would be one of Germany's first ever all-welded bridges. The total length was 442 meter, having a centre span of 236 meter and two end spans of 103 meter each (Fig. 3.58).

The cross-section of the bridge in mid-span had an 11.0 meter wide lower flange, a 29.5 meter wide upper flange, and a depth of 5.88 meter. The upper flange and the inclined web plates were in the longitudinal direction strengthened by bulb flat stiffeners (just as the case was for the Cleddau Bridge and the West Gate Bridge), while the lower flange had T-profiles (which also was used for the Britannia Bridge) (Fig. 3.59).

In order to minimize the obstruction to the boat traffic, one had chosen to cantilever out the centre span from both directions (just as for the Danube Bridge). The cantilever arm was about 100 meters long, and there was just the erection of the final section (an 18 meter high lift of a 16 meter long and 85 ton bridge element from the water level) as the cantilever arm collapsed approximately halfway out (Fig. 3.60) (compare also the photo in Fig. 3.61).

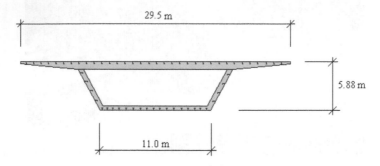

Figure 3.59 Cross-section of the Rhine Bridge, near Koblenz.

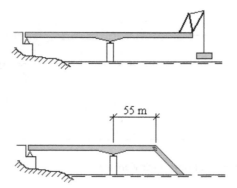

Figure 3.60 During erection of the main span (using the free cantilevering technique) the cantilever arm buckled approximately half-way out and fell into the river Rhine.

Figure 3.61 Photo taken just after the collapse (compare Fig. 3.60).

The negative bending moment that is the effect of self-weight, the weight of the crane, and the new girder element (plus the additional inertia forces as the cantilever arm must be assumed to have been set in swaying), created too much compression for the lower flange to carry, which buckled and unluckily transformed the statically determinate system into a mechanism. The strange thing, however, was that the buckle came halfway out, and not at the inner support, where the bending moment was the greatest. Several factors do influence here:

- The bridge was haunched over the supports, i.e. stronger at these locations due to larger depth.
- The position of the buckle – 55 meter out from the support – is approximately where one could expect the zero-moment points to be for the continuous system, and it can thus be assumed that the stiffeners there were "minimized" in size.
- Last, but not least, the buckle did come exactly where the longitudinal stiffeners in the lower flange were spliced (i.e. in the joint between two elements) – a splice that quite clearly proved to be insufficient, as we will see in the following.

When two elements were to be joined together, there was a horizontal gap of 225 millimeter from each edge, to allow for the automatic welding in the transverse direction of the elements (without interruption of the many stiffeners) (Fig. 3.62).

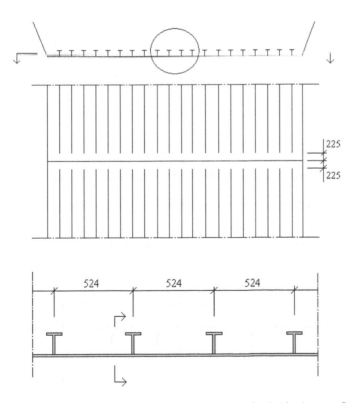

Figure 3.62 The longitudinal stiffener configuration of the box-girder bridge bottom flange.

As the longitudinal stiffeners were to be made intact again (i.e. continuous over the joint), the choice was to "hang" a T-profile on to the edges of two adjacent stiffeners, instead of filling the gap with an exact profile. The intention was (and quite rightly so, looking at the problem from one point of view) to avoid intersecting welds, that is a weakening factor concerning fatigue and brittle fracture strength (Fig. 3.63).

This choice of configuration created a free vertical distance of 25 millimeter between the inserted profile and the flange plate, which unfortunately did show to be the direct reason why buckling was initiated. Almost the same type of splice had been used at the West Gate Bridge (compare Fig. 3.49), however, there – in contrast to this bridge – was this weakening not the "triggering factor" (as it did show to be here).

We will in the following study and analyze the load-carrying capacity of the flange plate in this more or less unstiffened zone. We lack information about the thickness of the plate, and the yield strength of the steel material, however, as a qualified guess we assume $t = 10$ mm and $f_y = 360$ MPa. The flange plate is 11.0 meter wide and 450 millimeter long (in between the stiffener edges), and we assume simple supports along the four edges of the plate (Fig. 3.64).

We could perhaps assume the plate of being clamped to a certain degree close to the stiffener edges, however, as the stiffeners are widely separated (having a spacing of 524 mm) in relation to the distance in the longitudinal direction (i.e. 450 mm) we neglect this effect. Besides, as the free buckling length of the flange plate also is longer

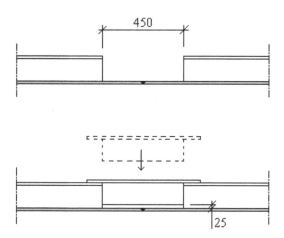

Figure 3.63 The unfortunate choice of splicing the longitudinal stiffeners.

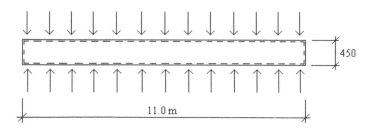

Figure 3.64 The unstiffened part of the bottom flange plate, in between the longitudinal stiffener ends.

(i.e. longer than 450 mm) in between the stiffeners, this compensates our assumption regarding the simple supports at a distance of 450 millimeter.

We calculate the critical buckling stress for the flange plate (Eqs. 3.7–3.8):

$$k = \left(\frac{11.0}{0.450} + \frac{0.450}{11.0} \right)^2 = \underline{599.5} \quad \text{(with } m = 1\text{)} \tag{3.7}$$

$$\sigma_{cr} = 599.5 \cdot \frac{\pi^2 \cdot 210000}{12 \cdot (1 - 0.3^2) \cdot \left(\frac{11.0}{0.010} \right)^2} = \underline{94.0 \, \text{MPa}} \tag{3.8}$$

Just to assure ourselves of the reasonableness of this value, we calculate the critical buckling stress, but based on the Euler theory (knowing that the difference in result given the theory for a plate and the theory of a strut should be small, as the loaded width is small (compare pages 6 and 11) (Eq. 3.9):

$$\sigma_{Euler} = \frac{\pi^2 \cdot 210000}{12 \cdot (1 - 0.3^2) \cdot \left(\frac{450}{10} \right)^2} = \underline{93.7 \, \text{MPa}} \tag{3.9}$$

The small difference in results – which confirmed the validity – is also on the right side. The critical buckling stress, based on the plate theory, will even for very wide plates be slightly higher, due to the double curvature at the edges.

We calculate the effective width according to the Eurocode, in order to receive the maximum load-carrying capacity in the ultimate loading state (compare pages 15–16) (Eq. 3.10):

$$\bar{\lambda}_p = \sqrt{\frac{360}{94.0}} = \underline{1.957} \quad \rho = \frac{(1.957 - 0.22)}{1.957^2} = \underline{0.454}$$

$$\Rightarrow \quad b_{eff} = 0.454 \cdot 11.0 = \underline{4.99 \, \text{m}} \tag{3.10}$$

Now we compare this value with the effective width for the stiffened flange plate on either side of the spliced zone (i.e. the effective width for an 11.0 meter wide stiffened plate having 20 stiffeners, spaced 524 millimeters). As it is now a much smaller free distance in the transverse direction, the critical buckling stress will increase markedly (Eqs. 3.11 & 3.12):

$$\sigma_{cr} = 4.0 \cdot \frac{\pi^2 \cdot 210000}{12 \cdot (1 - 0.3^2) \cdot \left(\frac{524}{10} \right)^2} = \underline{276.5 \, \text{MPa}} \tag{3.11}$$

$$\bar{\lambda}_p = \sqrt{\frac{360}{276.5}} = \underline{1.141} \quad \rho = \frac{(1.141 - 0.22)}{1.141^2} = \underline{0.707}$$

$$\Rightarrow \quad b_{eff} = 0.707 \cdot 0.524 \cdot 21 = \underline{7.78 \, \text{m}} \tag{3.12}$$

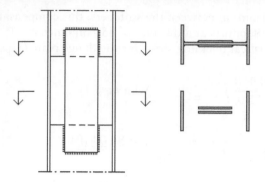

Figure 3.65 A simplified model that could be used to explain and to understand the behaviour of the weak splicing of the longitudinal stiffeners in the box-girder bridge bottom flange – an axially loaded column having the web spliced with two cover plates over a gap.

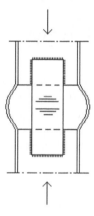

Figure 3.66 Local buckling of the column having the web spliced.

In relation to the stiffened parts on either side, the capacity has been reduced to only 64% in the spliced zone (4.99/7.78) – a major decrease, that to an almost certainty not was taken into consideration!

As the flange plate was not in close (and composite) contact with the longitudinal stiffeners in the spliced zone, it buckled and thus triggered the collapse of the entire cross-section. As a similitude one can consider a column, that has a very special splice where a part of the web is missing, and that the web has been replaced with two separate cover plates (Fig. 3.65).

For such a weak splice it is not difficult to imagine the scenario at the first heavy loading – the flange plate of the column, and the cover plates, will for sure buckle out (as the splice more or less also can be regarded as a hinge) (Fig. 3.66).

Something that in addition did weaken the splice detail in the Rhine Bridge was the inevitable welding deformations that more or less forced the flange plate into a deflected shape in between the stiffener edges, and this already from the start, i.e. before the structure came to be loaded (the deformation below is enlarged and only in principle) (Fig. 3.67).

Figure 3.67 Welding deformations in the transverse splice of the bottom flange plate tend to produce out-of-plane deformations.

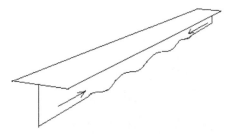

Figure 3.68 The free and unstiffened lower edge of the inserted T-profile (in the gap) is a weak part with respect to normal stress buckling.

An additional effect, that most certainly also not was considered, was the "splice profile" (i.e. the inserted T-profile), which had a lower flange that was free and unsupported, and this made it prone for normal stress buckling along this edge (just as the case was for the free flange edge of the West Gate Bridge) (Fig. 3.68).

After the collapse, the Rhine Bridge was re-built, now with the gap between the splice profile and the flange plate closed, and with extra strengthening of the bridge in general. For the erection of the new bridge, free cantilevering was once again used (despite the consequence for the original bridge), however now with a reduced maximum cantilever length – a 60-meter long mid-section was this time lifted up from two cantilever ends.

3.3.5 *The Zeulenroda Bridge*

Up until the end of the 1990s, the belief was that the four bridges in Wien, Milford Haven, Melbourne, and Koblenz, were the only major examples of box-girder bridge collapses during erection. Imagine the surprise as it in 1998 was reported in the German scientific journal Stahlbau of another box-girder bridge collapse – a collapse that also had taken place in the early 1970s. The East German authorities had kept this incident a secret to the Western world, and it was not until the archives were opened that the information became available for the general public. A probable reason why this was hidden away was the fact that the collapse happened at the same day as the anniversary of the Berlin Wall. Twelve year to the day, after the building of the Berlin Wall, and less than four years after the buckling of the Danube Bridge, the box-girder bridge in Zeulenroda (approximately 100 kilometers south of Leipzig, close to the border of the Czech Republic, then Czechoslovakia) did collapse – on the 13th of August 1973 – killing four people.

The bridge was to be a continuous box-girder bridge in six spans, having a total length of 362 meter. Just before the collapse, the second span had been cantilevered

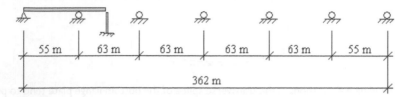

Figure 3.69 Elevation of the Zeulenroda Bridge during erection.

Figure 3.70 A longitudinal view of the Zeulenroda Bridge during erection.

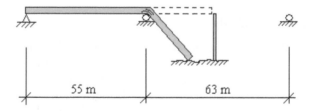

Figure 3.71 Collapse mechanism.

half-way out, and the next element was to be assembled, carried by a temporary support in mid-span (Fig. 3.69) (see also Fig. 3.70).

However, the temporary support in mid-span never came to be in use, as the bridge suddenly gave way and fell to the ground (Fig. 3.71).

There was no exact and detailed explanation put forward in the Stahlbau article of the reason for the collapse, but still we will in the following give it a try to analyze the bridge cross-section to see if we can, somehow, prove that the bridge was inadequate

Figure 3.72 The unexpected and sudden collapse of the Zeulenroda Bridge.

in its design. The structural drawing, showing the cross-section of the box-girder, is the following (given in Fig. 3.73).

This drawing gives us the information about the cross-section that we need for our analysis (Fig. 3.74).

The assembly of the bridge was, as has been mentioned previously, at the stage where the cantilever was halfway out, and that the next element was to be connected to the cantilever end, supported also by a temporary support (Fig. 3.75).

As the bridge gave way, the new element was not yet at the crane, instead it was positioned behind the same. The approximate positions of the crane and the element out on the cantilever were the following (Fig. 3.76).

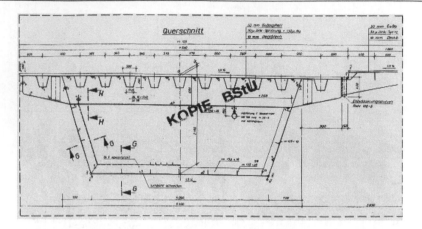

Figure 3.73 Cross-section of the Zeulenroda Bridge, as given in the structural drawings.

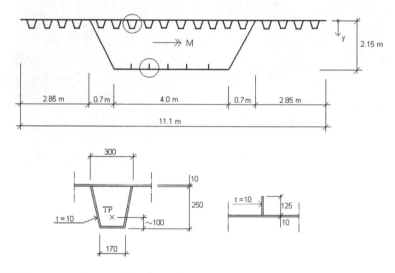

Figure 3.74 The cross-section used for the analysis to come in this text.

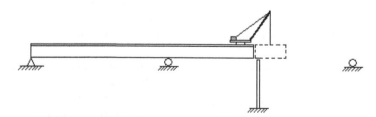

Figure 3.75 The next sequence in the assembly of the bridge (if it not had been for the collapse that is).

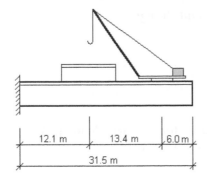

Figure 3.76 The position of the crane and the new element just prior to the collapse (compare Fig. 3.72).

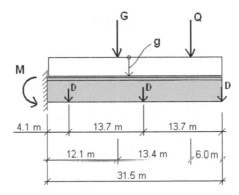

Figure 3.77 The different loads acting on the cantilever arm.

In order to get an idea of the loading action on the cantilever arm at this particular occasion, we need not only the positions of the different loads, but also the size of the same. The weight Q of the crane, we set an absolute minimum value to (Fig. 3.77).

$Q \geq 100 \, kN$

After analyzing the cross-section and finding the constants, we achieve the remaining loads on the cantilever. The new element is 13.7 meters long, and is without transverse cantilevers.

$G = 305.8 \, kN$ (the weight of the new element)

$g = 25.8 \, kN/m$ (self-weight per meter bridge)

$D = 5.4 \, kN$ (the weight of the diaphragm)

The cross-sectional constants being:

$$A = 287976 \text{ mm}^2$$
$$I_x = 19.362 \times 10^{10} \text{ mm}^4$$
$$y_{n.a.} = 574 \text{ mm}$$

The design bending moment for transient loading (i.e. temporary loading during construction) becomes (Eqs. 3.13 & 3.14):

$$M = 25.8 \cdot \frac{31.5^2}{2} + 5.4 \cdot (4.1 + 17.8 + 31.5)$$
$$+ 305.8 \cdot 12.1 + 100 \cdot (31.5 - 6) = \underline{19339\,\text{kNm}} \tag{3.13}$$

$$M_{sd} = \gamma_G \cdot M = 1.35 \cdot 19339 = \underline{26107\,\text{kNm}} \tag{3.14}$$

The maximum normal stresses in the upper and lower flanges become (Eqs. 3.15 & 3.16):

$$\sigma_{upper} = \frac{M_{sd}}{I_x} \cdot y_{n.a.} = \frac{26107 \cdot 10^{-3}}{19.362 \cdot 10^{-2}} \cdot 574 \cdot 10^{-3} = \underline{77.4\,\text{MPa}} \tag{3.15}$$

$$\sigma_{lower} = \frac{M_{sd}}{I_x} \cdot (h - y_{n.a.}) = \frac{26107 \cdot 10^{-3}}{19.362 \cdot 10^{-2}} \cdot (2150 - 574) \cdot 10^{-3}$$
$$= \underline{212.5\,\text{MPa}} \tag{3.16}$$

The lower flange, that shall carry a compressive stress of 212.5 MPa, is stiffened in the longitudinal direction by five flat steel bars, and in the transverse direction by a stiffener each 2.74-meter (Fig. 3.78).

The first step will be to check the maximum capacity of the flange plate in between the longitudinal stiffeners. We start by calculating the critical buckling stress of the

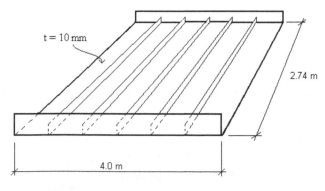

Figure 3.78 Transverse and longitudinal stiffeners in the lower flange.

same. We assume the free width of being equal to the centre distance between the stiffeners (Eqs. 3.17 & 3.18):

$$a = 2.74 \ m \quad b = \frac{4.0}{6} = 0.667 \ m$$

$$\frac{a}{b} = \frac{2.74}{0.667} = 4.1 \quad \Rightarrow \quad \underline{\underline{k = 4}} \tag{3.17}$$

$$\sigma_{cr} = 4 \cdot \frac{\pi^2 \cdot 210000}{12 \cdot (1 - 0.3^2) \cdot \left(\frac{667}{10}\right)^2} = \underline{\underline{170.6 \ \text{MPa}}} \tag{3.18}$$

Already here we see that something is not quite right – the actual compressive stress of 212.5 MPa is markedly exceeding the ideal critical buckling stress – and we can suspect that the post-critical reserve strength is not going to be enough. We will still carry out the analysis. First we check the cross-section class of the flange plate. The steel quality is St 38, which has a yield strength of 235 MPa (Eqs. 3.19 & 3.20):

$$\text{Class 3:} \quad \frac{b}{t} \leq 42 \cdot \sqrt{\frac{235}{f_y}} = \underline{\underline{42}} \tag{3.19}$$

$$\text{Actual slenderness:} \quad \frac{b}{t_f} = \frac{667}{10} = \underline{\underline{66.7}} \tag{3.20}$$

$$\Rightarrow \text{Class 4}$$

We continue by calculating the effective width (for a cross-section in Class 4) (Eqs. 3.21–3.23):

$$\overline{\lambda}_p = \sqrt{\frac{235}{170.6}} = \underline{\underline{1.173}} \tag{3.21}$$

$$\rho = \frac{(1.173 - 0.22)}{1.173^2} = \underline{\underline{0.693}} \tag{3.22}$$

$$\Rightarrow \quad b_{eff} = 0.693 \cdot 667 = \underline{\underline{462 \ \text{mm}}} \tag{3.23}$$

The load-carrying capacity – expressed as the axial force capacity – then becomes (Eq. 3.24):

$$N_{c.Rd} = \frac{A_{eff} \cdot f_y}{\gamma_{M1}}$$

$$= \frac{462 \cdot 10^{-3} \cdot 10 \cdot 10^{-3} \cdot 235 \cdot 10^3}{1.0} = \underline{\underline{1086 \ \text{kN}}} \tag{3.24}$$

which we compare to the design axial force (Eq. 3.25):

$$N_{c.Sd} = A_{tot} \cdot \sigma_{lower}$$

$$= 667 \cdot 10^{-3} \cdot 10 \cdot 10^{-3} \cdot 212.5 \cdot 10^3 = \underline{1417\,\text{kN}} \tag{3.25}$$

$\Rightarrow$ Insufficient load-carrying capacity!

If the flange plate (in between the stiffeners) would have had a sufficient load-carrying capacity, the next step in our analysis would have been focusing on the longitudinal stiffeners alone. For the completeness of our analysis we continue to do this check, and we assume the longitudinal stiffeners are having the dimension that was originally given in the cross-section, i.e. $125 \times 10\,\text{mm}^2$ – the thickness is, however, crossed out and adjusted to 20 millimeters (an adjustment made afterwards – as a correction of what it should have been?). The simplest model, but also the most practical one, is to regard the longitudinal stiffeners as separate units, without any support in the transverse direction from the flange plate. The stiffeners are then assumed to buckle independently of each other, in the out-of-plane direction relative the flange plate plane (in between the transverse cross-beams) as isolated elements (Fig. 3.79).

We have then the following cross-section to analyze (Fig. 3.80).

A check of the cross-section class of the longitudinal stiffener (Eqs. 3.26 & 3.27):

$$\text{Class 3:} \qquad \frac{c}{t} \leq 14 \cdot \sqrt{\frac{235}{f_y}} = \underline{14} \tag{3.26}$$

$$\text{Actual slenderness:} \qquad \frac{c}{t} = \frac{125}{10} = \underline{\underline{12,5}} \tag{3.27}$$

As the actual slenderness lies below the slenderness limit of Class 3, the longitudinal stiffener (being subjected to axial compression) has a full capacity to fully plastify. Note that this is only with respect to the local instability risk.

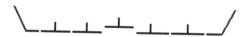

Figure 3.79 As a simplified approach the longitudinal stiffeners are assumed to buckle independently of each other.

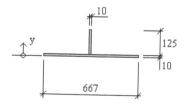

Figure 3.80 Cross-section of the longitudinal stiffener and the interacting part of the lower flange plate.

Cross-sectional constants:

$$A_{gross} = 7920 \text{ mm}^2$$

$$A_{eff} = 5870 \text{ mm}^2 \text{ (with the effective width as before)}$$

$$y_{n.a.} = 15.7 \text{ mm}$$

$$I_x = 6.48 \times 10^6 \text{ mm}^4$$

The design buckling resistance (Eqs. 3.28–3.38):

$$N_{b.Rd} = \frac{\chi \cdot \beta_A \cdot A \cdot f_y}{\gamma_{M1}} \tag{3.28}$$

$$i = \sqrt{\frac{I}{A}} = \sqrt{\frac{6.48 \cdot 10^6}{7920}} = \underline{28.6 \text{ mm}} \tag{3.29}$$

$$\lambda = \frac{l_c}{i} = \frac{2.74}{28.6 \cdot 10^{-3}} = \underline{95.8} \tag{3.30}$$

$$\lambda_1 = 93.9 \cdot \varepsilon = \underline{93.9} \tag{3.31}$$

$$\beta_A = \frac{A_{eff}}{A} = \frac{5870}{7920} = \underline{0.741} \tag{3.32}$$

$$\bar{\lambda} = \frac{\lambda}{\lambda_1} \cdot \sqrt{\beta_A} = \frac{95.8}{93.9} \cdot \sqrt{0.741} = \underline{0.878} \tag{3.33}$$

$$\text{buckling curve } c \text{ (welded)} \quad \Rightarrow \quad \chi = \underline{0.614} \tag{3.34}$$

$$N_{b.Rd} = \frac{0.614 \cdot 0.741 \cdot 7920 \cdot 10^{-6} \cdot 235 \cdot 10^3}{1.0} = \underline{847 \text{ kN}} \tag{3.35}$$

To be compared to the actual normal force that a longitudinal stiffener shall carry (we assume the normal stress to be evenly distributed over the cross-section) (Eq. 3.36):

$$N_{b.Sd} \approx 212.5 \cdot 10^3 \cdot 7920 \cdot 10^{-6} = \underline{1683 \text{ kN}} \tag{3.36}$$

From the results we see that the longitudinal stiffeners also they were given too small dimensions, so the bridge was quite clearly doomed to fail during erection. And even if consideration is made regarding a more refined global buckling model – where, for example, the influence of the stiffening effect in the transverse direction from the flange plate is taken into consideration (compare section 4.2, and example 3 and 7), and also, for the real case, accept a capacity calculated without safety factors on the load and yield strength – there are more reducing effects that have to be considered, which *lower* the load-carrying capacity. In order for a longitudinal stiffener to function as a rigid nodal point at buckling of the flange plate, it has not only to be sufficiently stiff in the out-of-plane direction (i.e. in the vertical direction relative the plane flange plate), but

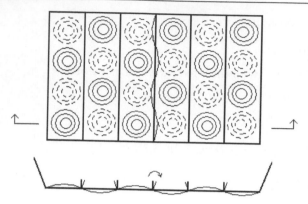

Figure 3.81 The sine wave buckling pattern in the flange plate has the effect that they tend to rotate the longitudinal stiffeners in a sine wave pattern as well.

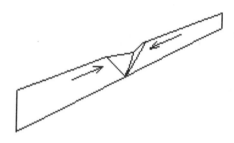

Figure 3.82 Possible "folding" mechanism of a longitudinal stiffener due to a local damage.

also be torsionally stiff, in order not to deform in the transverse direction due to the plate buckling (Fig. 3.81).

Flat steel bars as stiffeners are "torsionally soft", which make them susceptible for welding deformations, and this increases the tendency even more for lateral deformations. In addition, there is also the probability of hits and damages during transport, handling and assembly. A local damage can make the stiffener lose most of its load-carrying capacity with respect to the global buckling resistance (Fig. 3.82).

And if one longitudinal stiffener is eliminated, it also means that the total load-carrying capacity of the flange plate is reduced (Fig. 3.83) (as the effective width of the plate decreases).

Thus, the advice is that you never should choose flat bars as longitudinal stiffeners (as was the case here in the Zeulenroda Bridge, and also in the Danube Bridge), because of the above-mentioned reasons. They have not enough bending stiffness (and rotational stiffness) in the transverse direction. Profiles having a large bending stiffness in the transverse direction (besides the vertical direction) are for example L-profiles or T-bars (Fig. 3.84).

There are also more advantages than have been discussed above using these kinds of profiles in comparison to flat steel bars. A flat steel plate has an outstand flange edge, that is, even if it is straight and without imperfections, very sensitive for normal stress buckling (compare the discussion regarding the West Gate Bridge, and the free flange

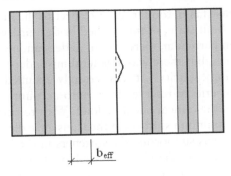

Figure 3.83 The effect on the overall ultimate loading capacity because of a damaged stiffener.

Figure 3.84 A more desirable choice of stiffener types.

edge). The relatively higher torsional stiffness of L- or T-profiles also gives a higher clamped condition for the flange plate edges (the transverse rotation is prevented to a higher degree – compare Figs. 3.81 and 3.84), which increases the critical buckling stress of the flange plate.

As a final comment regarding the choice of longitudinal stiffener cross-section for the Zeulenroda Bridge, one can say that the wrong profile was chosen, and this in combination with an insufficient load-carrying capacity of the bridge as a whole. Besides a thicker flange plate, and more rigid stiffeners, it would also have been motivated with a greater depth of the bridge cross-section that would have lowered the bending stresses. The Zeulenroda Bridge was very shallow in relation to the span length (compare for example with the Cleddau Bridge).

3.4 Summary

We have in section 3.2 (The Britannia Bridge) seen how the buckling risk could be treated with great success, and this without having full knowledge of the theories, however, compensating this with tests, and with a great will of learning and understanding. In section 3.3, it was shown a number of collapses in modern time, where the risk of buckling – in contrast to the Britannia Bridge – not was treated with proper care, and this despite the fact that they more or less had full access to both theory and design methods. However, in the end of the 1960s it was still so that the codes did not give full support to the engineers, as one would have wished, instead they had to design the box-girder bridges much up to their own ability. The concept of box-girder bridges was new and exciting, and the engineers did not see any limitations, only opportunities. The free cantilever erection method was just such an opportunity that was used – it was here given the chance of an easy and fast erection method, which did not, for example, obstruct the boat traffic on a river. This particular concept was, however, in twice the sense taken too far. With respect to the maximum load-carrying capacity in the ultimate limit state, of thin-walled bridge girder cross-sections in general, and slender plates in particular, there was an understanding of the existence of the post-critical reserve effects, but there was a general misunderstanding about the final and ultimate load-carrying capacity – it was wrongly assumed that the final loading capacity (taking the post-critical reserve effects into account) should approximately be three times as large as the critical buckling load. With respect to the necessity that stiffeners should have a minimum stiffness in the transverse direction, the general idea was that the flange plate was supporting the stiffeners in the lateral direction, so this requirement needs not to be considered (which we have seen is definitely not the case).

Each failure case has, however, contributed to the understanding of the buckling phenomenon in thin-walled and slender plates, and given us its individual clue.

The Danube Bridge

Due to the chosen erection method (free cantilevering), the statically indeterminate system was bent (i.e. strained) to such a level, that the additional constraint forces coming from the temperature drop released major buckling of the bridge in two locations. Even if one has to assume the stiffeners being strong enough for the in-service loading, they were demonstrably not strong enough to meet the temporary loading during erection.

The Cleddau Bridge

The knowledge concerning how to design and stiffen a diaphragm for concentrated loads during erection was in this case definitely inadequate.

The West Gate Bridge

Unfortunate choice of erection technique, in combination with a complete ignorance regarding the primary statics of a simply supported girder.

The Rhine Bridge

A badly chosen stiffener detail over the splice between two bridge elements induced local buckling that, as a consequence, made the entire bridge collapse.

The Zeulenroda Bridge

In this particular failure case, one could definitely raise the question regarding the design of this bridge – was it up to the contractor to chose the erection method, without a proper check by the designer?

Chapter 4

I-girders

4.1 Introduction

In this chapter we will concentrate on the background of the design of I-girders with respect to buckling, such as normal stress buckling, local buckling under concentrated loads, and shear buckling. The reason why we study I-girders, instead of continuing with box-girder bridges more in detail, is that an I-girder, as a structural element, is less complex, more "two-dimensional". Even if the bending theory is applicable for both bridge types, there are special effects that have to be considered in box-girder bridges, e.g. torsion and shear lag. The last-mentioned effect is the influence that wide plates have on the normal stress distribution – the flexibility in the central parts makes the stresses concentrate more to the stiff edges of the plate (i.e. to the web plate connection). The shear lag effect is especially large at wide plates in short spans, and influences the most in the serviceability limit state where consideration has to be made on the bending stiffness distribution in statically indeterminate systems (and hence the moment distribution of the same). In the ultimate limit state, however, the plastic stress redistribution in the cross-section makes the shear lag effect more or less negligible.

In an I-girder bridge there are normally no wide flanges, and therefore no effect of shear lag. However, for I-girder bridges in composite action with a wide concrete deck, this effect is present in the behaviour (Fig. 4.1).

The concrete deck will interact with the steel girders and thus contribute to the capacity of the upper flange, and even if this particular type of bridge is more simple and traditional than a box-girder bridge, there is also flexibility in the concrete deck against shear forces that makes shear lag needed to be considered. Besides using an effective width of the concrete deck (taking shear lag into account), there are additional time-dependent effects that also have to be the considered (creep and shrinkage). We will in the following, however, focus on the more pure I-girder, having no composite action

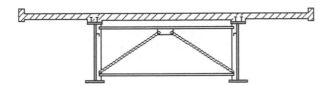

Figure 4.1 The cross-section of an I-girder bridge, having the concrete deck in composite action together with the steel girders.

coming from an interacting concrete deck, and this in order to make the buckling theories be displayed as clear as possible. As a final comment it is worth mentioning that the buckling theories are also valid for box-girder bridges, but with the exception that the shear buckling theory, which is presented in section 4.4, is not directly applicable for cross-sections having wide and flexible flange plates.

4.2 Normal stress buckling

For an I-girder subjected to a positive bending moment, there is a triangular stress distribution over the depth of the cross-section, with compressive stresses in the upper half, and tensile stresses in the lower – and the stresses are parallel to the normal (longitudinal) axis of the girder. The simply supported I-girder below is subjected to a uniform bending moment, which compresses the girder in the upper half, and elongates the same in the lower half (Fig. 4.2).

We also notice that the elongation in the lower part of the girder – despite the deflection – makes the roller bearing be displaced outwards in relation to the unloaded position (Fig. 4.3).

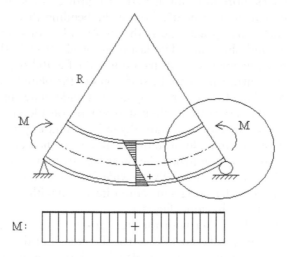

Figure 4.2 A simply supported I-girder subjected to uniform bending deformations.

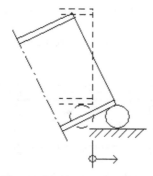

Figure 4.3 Elongation of the lower flange makes the roller bearing move outwards.

In the upper half, which is subjected to compression, there is a risk of three different kinds of instability phenomena because of the normal compressive forces (Figs. 4.4–4.6) (Eqs. 4.1, 4.2):

$$\sigma_{cr} = 23.9 \cdot \frac{\pi^2 \cdot E}{12 \cdot (1 - \upsilon^2) \cdot \left(\frac{b}{t}\right)^2} \tag{4.1}$$

$$\sigma_{cr} = 0.425 \cdot \frac{\pi^2 \cdot E}{12 \cdot (1 - \upsilon^2) \cdot \left(\frac{b}{t}\right)^2} \tag{4.2}$$

This last-mentioned instability phenomenon – lateral/torsional buckling – is a global buckling phenomenon, where the compression flange is buckling in the lateral (sideways) direction, followed by a rotation of the girder. There exist no post-critical reserve effects for this type of buckling, so this instability phenomenon is equal to the collapse of the entire system after lateral/torsional buckling has occurred. We will in the following, however, concentrate on the local buckling phenomena, which possess an ability of additional loading after buckling has occurred (according to the theory for plates).

If we consider a deep and slender I-girder, where both the flanges and the web are made of thin-walled plates, and thus belong to cross-section Class 4, the effective width, for the determination of maximum load-carrying capacity in the ultimate limit state (for a positive bending moment), will look like this (Fig. 4.7).

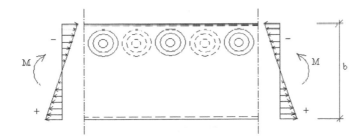

Figure 4.4 Possible normal stress buckling in the upper part of the web (web buckling). The critical buckling stress according to Eq. 4.1.

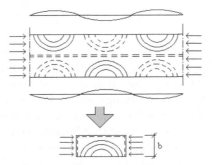

Figure 4.5 Possible normal stress buckling in the upper flange (flange buckling). The critical buckling stress according to Eq. 4.2.

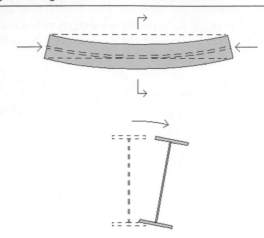

Figure 4.6 Possible out-of-plane (global) buckling of the upper flange in the lateral direction (lateral/torsional buckling).

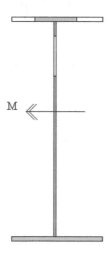

Figure 4.7 Effective net cross-section of a class 4 I-girder.

In order to increase the maximum load-carrying capacity, we could attach longitudinal stiffeners in those zones where the normal stress buckles are expected to have the largest amplitude (Fig. 4.8).

Stiffener type *a* is used for web buckling, type *b* for flange buckling, and type *c* for flange and web buckling combined. The latter type (which is perhaps most used for overhead cranes), has the advantage that it also stiffens against lateral/torsional buckling, as it – besides increases the flange area – also to a certain extent increases the rotational stiffness. Stiffener type *b* is also not a common choice for bridges, as the flanges there are normally made so compact that buckling of the same never should be a problem – a flange of an I-girder bridge is the main component with respect to bending moment resistance, and should therefore never be a reducing factor with respect to the normal stresses on the compressive side. If it is, from this point of view, advantageous to choose a flange plate having a low slenderness ratio *b/t* – i.e. small width and/or

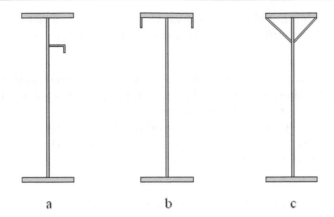

Figure 4.8 Different ways of increasing the buckling resistance of a slender and thin-walled I-girder.

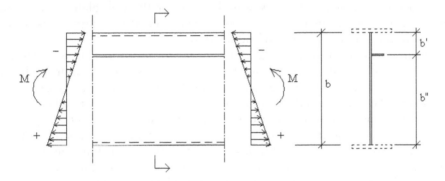

Figure 4.9 A longitudinal stiffener is positioned where web buckling is to be expected (for an unstiffened web). The upper and lower part of the web – above and below the stiffener – is then the web parts governing the load-carrying capacity (of the stiffened cross-section).

large thickness – it is the opposite with respect to the lateral/torsional buckling risk. In order for an I-girder to have a high safety against lateral/torsional buckling a wide flange is required, which can be hard to combine with the requirement regarding the local buckling risk. You have to balance the demands against each other. Despite that the risk of lateral/torsional buckling was negligible for the West Gate Bridge (due to the wide upper flange), they still forgot about the local buckling risk of the temporary free upper flange edge (as the slenderness ratio b/t was too high).

Stiffener type a – which is the normal type to use in a deep and slender (i.e. thin-walled) I-girder web – does contribute to the load-carrying capacity, not only because it minimizes the risk of buckling, but also that it increases the cross-sectional area as a small extra flange. The longitudinal stiffener is positioned where the maximum amplitude of the buckles is expected, which is at the level of approximately 20% of the depth of the web (from the compression flange). The load-carrying capacity of the stiffened cross-section is then governed by the risk of normal stress buckling on either side of the stiffener (that is assumed to act as a rigid nodal line) (Fig. 4.9).

Which part of the web plate (on either side of the stiffener) that is going to be governing with respect to the normal stress buckling risk, is decided by the critical buckling stress of the same. It is not always that clear which part that will be the governing one (with respect to the overall load-carrying capacity of the I-girder profile that is):

- For the panel above the stiffener, the loaded width (b') is small, and this will give a high critical buckling stress value – however, this is counteracted by a low buckling coefficient (k), due to dominating compression.
- For the panel below the stiffener, the loaded width (b'') is large, and this will give a low critical buckling stress value – however, this is counteracted by a high buckling coefficient, due to dominating tension.

Normally it is so that buckling of the unstiffened plate panels on either side of the stiffener is not limiting the load-carrying capacity, at least not for normal sized girders. At extra large I-girders there is often, however, a need for more than one longitudinal stiffener in the web, just as we saw was the case for the web plates in box-girder bridges. Multiple longitudinal stiffeners can also be motivated if extra robustness and stability is required during transport and handling.

The requirement that a longitudinal stiffener shall be a fix nodal line with respect to web buckling, calls for the need that the stiffener not only should be compact, but also sufficiently stiff so that the stiffener does not buckle in the out-of-plane direction of the web plate. There are two things governing this resistance against global buckling; besides the bending stiffness of the stiffener (EI_{st}), it is also the buckling length of the same. The governing parameter with respect to the buckling length, is the distance a between the vertical stiffeners (that has, among other things, the function of minimizing the buckling length of the longitudinal stiffener). When a longitudinal stiffener is buckling in the lateral direction (i.e. in the out-of-plane direction of the web plate), the web plate is also holding the stiffener back by its plate (bending) stiffness. The plate stiffness – which is a function of the web plate thickness t_w – is also a parameter that influences the buckling length of the longitudinal stiffener, and thus the buckling resistance of the same. A spring bed model is used to describe this stiffening effect in the transverse (out-of-plane) direction from the web plate (Fig. 4.10).

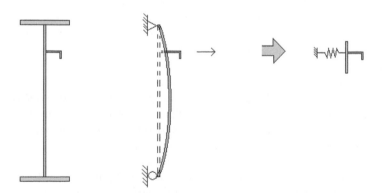

Figure 4.10 Buckling model for the longitudinal stiffener (interacting with the web plate).

The buckling length in the longitudinal direction (of the stiffener) can be calculated according to the following expression (Eq. 4.3):

$$a_c = 4.33 \cdot \sqrt[4]{\frac{I_{st} \cdot (b')^2 \cdot (b'')^2}{(t_w)^3 \cdot b}} \tag{4.3}$$

Depending on if the calculated buckling length, a_c, is smaller or larger than the distance a between the vertical stiffeners, the critical buckling load becomes (Eqs. 4.4, 4.5):

when $a_c \leq a$:

$$N_{cr} = 1.05 \cdot E \cdot \frac{\sqrt{I_{st} \cdot (t_w)^3 \cdot b}}{b' \cdot b''} \tag{4.4}$$

when $a_c > a$:

$$N_{cr} = \frac{\pi^2 \cdot EI_{st}}{a^2} + \frac{E \cdot (t_w)^3 \cdot b \cdot a^2}{35.7 \cdot (b')^2 \cdot (b'')^2} \tag{4.5}$$

The transverse, vertical stiffeners, that keep the buckling length down of the longitudinal stiffener, are designed with respect to the demand that the in-plane deflection of the same (i.e. out-of-plane relative to the web plate) is limited to $b/500$ for a balancing force, coming from a possible maximum inclination of the longitudinal stiffeners on either side of the vertical stiffener (normally this value is set to 2% of the longitudinal stiffener normal force) (Fig. 4.11).

With respect to the bending moment resistance of the cross-section the vertical stiff eners do not contribute to the load-carrying capacity (more than the reduction in

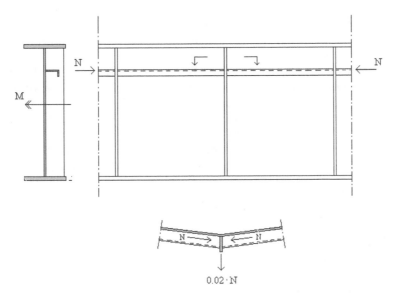

Figure 4.11 The out-of-straightness of the longitudinal stiffener parts induce a lateral force, which has to be balanced by the transverse stiffeners.

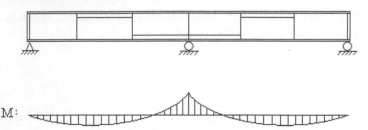

Figure 4.12 The longitudinal stiffeners in the different panels are positioned according to the bending moment distribution.

buckling length of the longitudinal stiffener they gave). With respect to the lateral/torsional buckling risk, the vertical stiffeners increase the stability somewhat though.

For continuous girders the longitudinal stiffener is positioned in those zones where the girder is subjected to normal force compression from the bending moment, i.e. alternating in the upper or lower half (Fig. 4.12).

4.3 Concentrated loads

In positions where a slender girder is subjected to a concentrated transverse load, there is always a risk that the load-carrying capacity of the girder is reduced with respect to the local buckling risk. This particular risk was already observed in the 1850s, when Houbotte from Belgium carried out a series of tests on simply supported girders, loaded in mid-span with a concentrated load. All of his tests ended in the same manner; the web plate did buckle in the upper part close to the point load. Depending on the distance to the girder edge, and if the web was stiffened or not at the supports, the buckling could as well have occurred in those areas (however, then in the lower half of the girder). Despite that the material used in steel girders today is stronger, and that the production is carried out with much higher precision and tolerance demands, the phenomenon is still the same, as it is first and foremost a matter of instability than a matter of strength (Fig. 4.13).

If the web plate is sufficiently compact (read: has a sufficiently small slenderness ratio d/t), the problem is reduced to the maximum possible contact pressure strength $R_{y \cdot Rd}$ of the web plate, which is governed by local yielding with successive crushing (plastic collapse) of the web plate. In the load distribution model a dispersion of the load – from a stiff bearing (on top of the flange), having the length s_s – is assumed transmitted down into the web (Fig. 4.14), (Eq. 4.6):

The additional dispersion length s_y is determined according to the expression:

$$s_y = 2 \cdot t_f \cdot \sqrt{\frac{b_f}{t_w}} \cdot \sqrt{\frac{f_{yf}}{f_{yw}}} \cdot \sqrt{1 - \left(\frac{\sigma_{f \cdot Ed}}{f_{yf}}\right)^2} \qquad (4.6)$$

t_f thickness of the flange
b_f width of the flange
t_w thickness of the web
f_{yf} yield strength of the flange
f_{yw} yield strength of the web
$\sigma_{f \cdot Ed}$ longitudinal stress in the flange

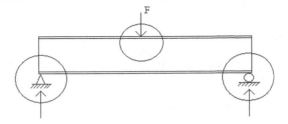

Figure 4.13 Possible locations for local buckling due to concentrated loads in an I-girder.

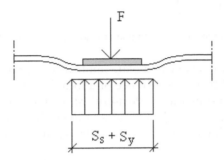

Figure 4.14 Local crushing (plastic collapse) in an I-girder having a compact web plate.

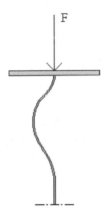

Figure 4.15 Crippling (local buckling) in an I-girder having a slender web plate.

The design crushing resistance is then obtained from the expression (Eq. 4.7):

$$R_{y \cdot Rd} = \frac{(s_s + s_y) \cdot t_w \cdot f_{yw}}{\gamma_{M1}} \qquad (4.7)$$

However, if the web plate is not being compact, but slender (i.e. with a high slenderness ratio d/t), the capacity will then instead be governed with respect to crippling (local buckling) (Fig. 4.15).

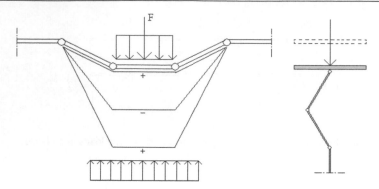

Figure 4.16 Buckling model in the ultimate limit state (taking post-critical reserve effects into account).

Just as the case was for buckling of a plate (uniform compression), the load-carrying capacity is not restricted by the occurrence of a local buckle in the web plate located directly under the concentrated load. Post-critical reserve effects – here in the shape of plastic hinges in the flange, and yield lines in the web (read: plastic folding of the web) – will give an additional strength to be added to the elastic buckling strength (as the one in Fig. 4.16).

Over the years it has been presented a large variety of different expressions explaining the ultimate load-carrying resistance, and these have more or less been semi-empirical, i.e. based on theoretical models (as the one in the figure above) in combination with observations from tests carried out. The expression presented in the first edition of the prestandard (ENV 1993-1-1) for the design crippling resistance was (Eq. 4.8):

$$R_{a \cdot Rd} = \frac{0.5 \cdot t_w^2 \cdot \sqrt{E \cdot f_{yw}} \cdot \left(\sqrt{\frac{t_f}{t_w}} + 3 \cdot \frac{t_w}{t_f} \cdot \frac{s_s}{d} \right)}{\gamma_{M1}} \tag{4.8}$$

When a member is also subjected to bending moments (which is adding longitudinal normal stresses to the zone affected by a concentrated load) the following criteria should also be satisfied (Eq. 4.9):

$$\frac{F_{sd}}{R_{a \cdot Rd}} + \frac{M_{sd}}{M_{c \cdot Rd}} \leq 1,5 \tag{4.9}$$

There are no distinct limits for the slenderness of the web plate when either local yielding (crushing) or local buckling (crippling) becomes governing, instead both capacities have to be checked (and the lowest will of course give the design resistance).

If the capacity with respect to a concentrated load was restricted to either local yielding or local buckling for girders subjected to bending (i.e. where the concentrated load is transferred to the supports by shear forces), it is another instability criterion that has to be checked when the load is positioned directly over a support (i.e. when no shear forces are present) (Fig. 4.17).

The web plate will in this case behave more or less as an "Euler strut", having a global buckle over the entire depth of the web (some restraining effect coming from flanges though) (Fig. 4.18).

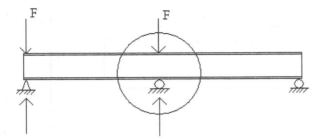

Figure 4.17 Concentrated loads that do not produce shear.

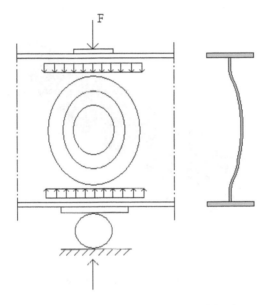

Figure 4.18 Overall buckling in the web due to a concentrated load transmitted directly through the web.

The design buckling resistance is calculated for a compression member, with an effective breadth at mid-depth equal to (Eq. 4.10):

$$b_{eff} = \sqrt{h^2 + s_s^2} \tag{4.10}$$

At the value of the stiff bearing length (s_s) equal to zero, the effective breadth will be equal to the depth h of the girder, i.e. with a distribution of 45° from the load application point to mid-depth. At increasing value of s_s, the spreading will, however, decrease.

An instability phenomenon that resembles this global buckling behaviour – but is not caused by a concentrated load – are girders with deep and slender webs being subjected to bending from an evenly distributed load. A FE-analysis of such a girder would show global buckling of the entire web plate. We do take the opportunity here to study a girder, having the shape according to Fig. 4.19.

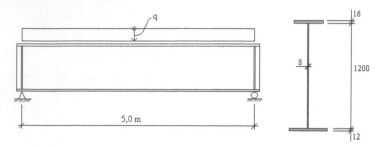

Figure 4.19   Elevation and cross-section of an I-girder used in the analysis to follow.

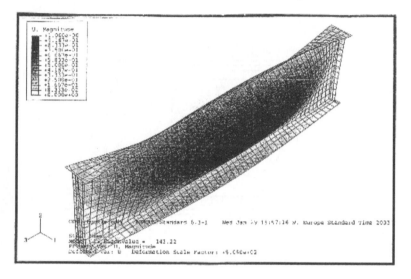

Figure 4.20   FE-analysis of the I-girder, showing the global buckling of the web.

The critical value of the evenly distributed load (q_{cr}) – with respect to global buckling of the web – will, according to a FE-analysis, be 143.2 kN/m, having a buckling mode shown in Fig. 4.20.

The global and extended web buckle in the major part of the span will have a wave length (read: buckling length) in the vertical direction equal to 0.6–0.7·d (Fig. 4.21).

We study a unit width of 1.0 meter in the longitudinal direction of the web plate, and calculate the critical buckling load according to the Euler theory (Fig. 4.22), (Eq. 4.11):

$$q_{cr} = \frac{\pi^2 \cdot EI}{L_{cr}^2} \cdot \frac{1}{(1 - v^2)}$$

$$\geq \frac{\pi^2 \cdot 2.1 \cdot 10^8 \cdot \left(\frac{1000 \cdot 8^3}{12}\right) \cdot 10^{-12}}{(0.7 \cdot 1200 \cdot 10^{-3})^2} \cdot \frac{1}{(1 - 0.3^2)}$$

$$= \underline{137.7 \, \text{kN/m}} \tag{4.11}$$

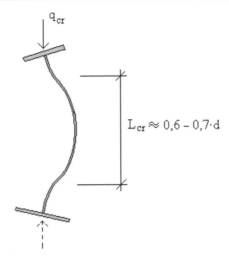

Figure 4.21 The free buckling length, L_{cr}, of the web.

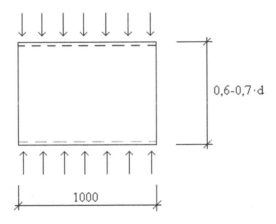

Figure 4.22 The critical buckling load is found by studying a unit width of 1.0 meter.

A result that correlates very well with the result of the FE-analysis!

In the current Eurocode (EN 1993-1-5: 2006) the design resistance, F_{Rd}, to local buckling under a transverse load in between two supports of a girder without longitudinal stiffeners is given to (Eqs. 4.12–4.20):

$$F_{Rd} = \frac{f_{yw} \cdot L_{eff} \cdot t_w}{\gamma_{M1}} \qquad (4.12)$$

f_{yw} yield strength of the web
L_{eff} effective length

$$L_{eff} = \chi_F \cdot l_y \qquad (4.13)$$

χ_F reduction factor

$$\chi_F = 0.5/\bar{\lambda}_F \ (\leq 1.0) \tag{4.14}$$

$\bar{\lambda}_F$ slenderness parameter

$$\bar{\lambda}_F = \sqrt{(l_y \cdot t_w \cdot f_{yw})/F_{cr}} \tag{4.15}$$

l_y effective loaded length ($\leq a$)

$$l_y = s_s + 2 \cdot t_f \cdot (1 + \sqrt{m_1 + m_2}) \tag{4.16}$$

s_s length of stiff bearing
t_f thickness of the flange

$$m_1 \quad (f_{yf} \cdot b_f)/(f_{yw} \cdot t_w) \tag{4.17}$$

f_{yf} yield strength of the flange
b_f width of the flange
t_w thickness of the web

$$m_2 \quad 0.02 \cdot (h_w/t_f)^2 \tag{4.18}$$

(if $\bar{\lambda}_f \leq 0.5$ then $m_2 = 0$)

h_w depth of the web
F_{cr} critical buckling load

$$F_{cr} = 0.9 \cdot k_F \cdot E \cdot t_w^3/h_w \tag{4.19}$$

k_F buckling coefficient

$$k_F = 6 + 2 \cdot (h_w/a)^2 \tag{4.20}$$

a distance between vertical stiffeners
E modulus of elasticity
γ_{M1} partial factor (instability)

The combined effect of a transverse load (on a compression flange) and bending moment has to be verified by the following interaction criterion (Eq. 4.21):

$$\frac{F_{Ed}}{F_{Rd}} + 0.8 \cdot \frac{M_{Ed}}{M_{pl,Rd}} \leq 1.4 \tag{4.21}$$

where:
$M_{pl,Rd}$ design plastic resistance of the cross-section consisting of the effective area of the flanges and the fully effective web irrespective of its section class.

Even if the normal procedure in bridge engineering is to attach vertical web stiffeners wherever concentrated loads are present (despite calculated design resistance of the

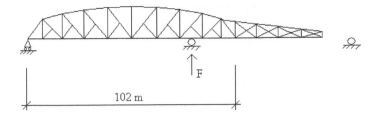

Figure 4.23 The launching procedure in 2003 of a truss bridge in Mexico.

Figure 4.24 Local buckling directly over the launching pad in the lower part of the web.

unstiffened web), it is in some cases not practical. When the chosen erection method is launching of the entire bridge over a sliding pad, the reaction force *F* will continuously move along the lower part of the web plate. This web plate will thus be subjected to a concentrated load in every position, which is a potential (local) buckling risk during the entire erection phase. As a 102 meter long truss bridge was to be launched out into its final position in 2003 in Mexico, they experienced this particular problem (Fig. 4.23).

The bridge was, during the erection, temporarily elongated with a so-called launching nose in order unload the truss as soon as the nose bridge reached to the other end. To a beginning the launching proceeded without problems; however, all of a sudden a buckle did appear in the lower half of the web plate of the lower chord. The actual weight on the launching pad was at the time approximately 500 tons. In the photo above (Fig 4.24) the buckle is seen in the lower region of the chord, over the right hand part of the launching pad. The upper part of the chord is partly concealed by a secondary longitudinal beam (the one having studs attached to its upper flange).

The lower chord of the truss had a closed box section, with three web plates (Fig. 4.25).

This box-girder profile had to – besides carry the concentrated load of 500 tons – also carry the additional axial normal force (compression) from the cantilevering action. In addition, there was also a local bending moment due to the transverse loading in between the supports (read: the joints of the truss) (Fig. 4.26).

Figure 4.25   Cross-section of the lower chord.

Figure 4.26   Local bending moments in the lower chord due to the introduction of the concentrated load (from the launching pad) in between the joints of the truss.

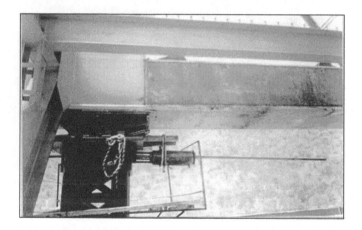

Figure 4.27   Extra cover plates were added to the exterior web plates of the lower chord.

All of these effects taken together became too much for the vertical web plates to carry, which buckled under the combined compression from concentrated load, normal axial force, and local bending moment.

After necessary steps were taken to restore the web plates into their original state (either by mechanical straightening or flame straightening), the girder was strengthened by adding an extra cover plate to the exterior webs. The launching process was then again resumed, now without any further problems (Fig. 4.27).

If this truss bridge in Mexico was an example of local buckling (crippling) of a web subjected to a concentrated load, the 2nd Narrows Bridge in Vancouver, Canada, was an example of global web buckling. This particular bridge, that was a continuous truss bridge, collapsed in 1958 during construction. The reason for the collapse was buckling of the web of a temporary girder that was not an actual part of the structure,

Figure 4.28 The support of the temporary vertical erection bent used for the 2nd Narrows Bridge.

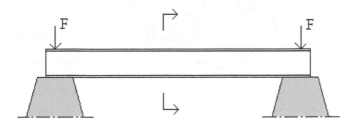

Figure 4.29 Elevation of the secondary girder supporting the erection bent.

instead this secondary girder had the function to carry a vertical erection bent (read: temporary support) for the cantilevering truss span during erection (Fig. 4.28).

This erection bent had the function to temporarily unload the superstructure by decreasing the maximum span length of the cantilever, and the secondary girder below supported this erection bent (Fig. 4.29).

The girder was a rolled standard profile (36WF160), and had a cross-section according to Fig. 4.30.

As the girder was not stiffened in the load application points (i.e. in the two positions directly over its supports), the web buckled when the loading from above became too

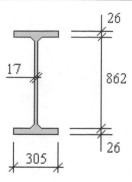

Figure 4.30 Cross-section of the secondary girder.

Figure 4.31 As the web of the secondary girder buckled the erection bent failed and the supported truss above collapsed.

much for the web to carry. As the web of this secondary (and temporary) member did buckle, it also meant the entire collapse for the cantilevering truss, as well as for the span lying behind (Fig. 4.31).

Too much of the concentration was focused up above on the huge trusses, that they forgot to stiffen the web of the "negligible" girder down below. A couple of simple web stiffeners, on both sides of the web at the load application points, should have avoided this collapse (Fig. 4.32).

The Eurocode describes how to design vertical web stiffeners. The stiffeners, and a part of the web plate, are considered as a homogeneous compression member that is checked with respect to global buckling in the transverse lateral direction, i.e. in the out-of-plane direction relative the web plate (Fig. 4.33).

The buckling length of this compression member shall be chosen with respect to the lateral and rotational restraint at the flanges, however, not less than $0.75 \cdot d$. Besides

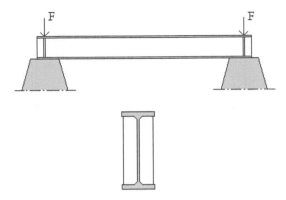

Figure 4.32 Some small and simple stiffeners in the secondary girder web would have prevented the collapse.

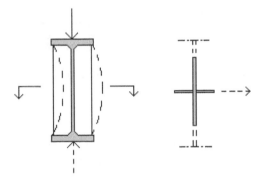

Figure 4.33 Out-of-plane buckling model of a stiffened web plate.

checking the global buckling resistance of the member, local buckling (crippling) and local yielding (crushing) should also be checked.

4.4 Shear buckling

A simply supported girder is a statically determinate structure, however, with respect to the inner mode of action, regarding how the load is transferred by shear, the system is statically indeterminate. And just as the case was for normal stress buckling (and in a way also for concentrated loads), the girder has a post-critical reserve strength that enables for additional loading after that shear buckling has occurred. The ability to redistribute load, and finding alternative load-paths, is especially apparent, as it is connected to an intuitive understanding of the behaviour of structures in general. We can start by studying a simply supported I-girder having three web panels (the areas in between the vertical stiffeners), symmetrically loaded with two equally large point loads (Fig. 4.34).

For the continued discussion we transform this I-girder into an equivalent truss (statically determinate, however, with respect to the inner system, statically indeterminate) (Fig. 4.35).

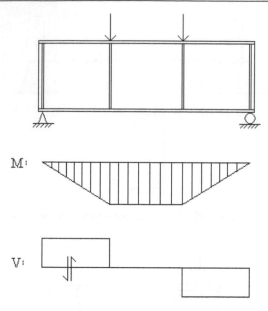

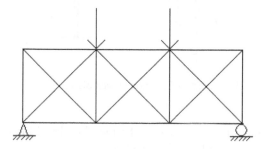

Figure 4.34 Moment and shear distribution of a simply supported I-girder having two symmetrically positioned, and equally large point loads.

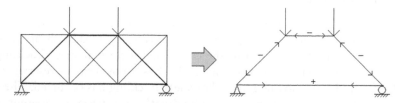

Figure 4.35 The I-girder transformed into an equivalent truss.

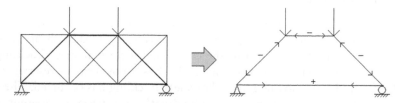

Figure 4.36 A Queen Post truss system having dominating compression.

From this equivalent truss, we could distinguish two separate Queen Post trusses – one having dominating compression (Fig. 4.36), and a reversed (mirror-wise), having dominating tension (Fig. 4.37).

Due to the symmetrical loading, the diagonals in the mid panel become inactive (also check the shear force diagram in Fig. 4.34).

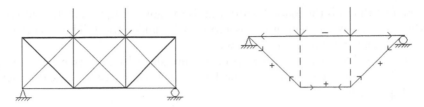

Figure 4.37 A Queen Post truss system having dominating tension.

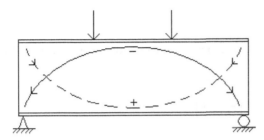

Figure 4.38 The compression and tension flow in an I-girder is similar to that of an equivalent truss.

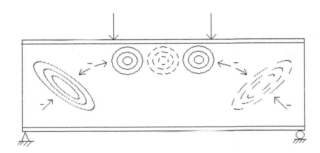

Figure 4.39 Along the compression flow there is a potential risk of different kinds of buckling to occur.

To return to the simply supported I-girder that we started with, so is the function for the same (with or without vertical stiffeners) principally a combination of the two separate Queen Post trusses (Fig. 4.38).

The fact is that the I-girder web is even more similar in action to a truss (i.e. with diagonals in tension and compression) having the vertical stiffeners in place.

With respect to instability, there is a potential buckling risk all along the compression "parabola", just as the case was for the up-right standing Queen Post truss, where every member in compression also could buckle. Between the point loads (in the area where the flanges were having a constant normal force) we find the normal stress buckles in the upper half of the web (compare section 4.2), and closer to the supports – where the shear force has its maximum – we find diagonal shear buckles. Both buckling types have occurred due to an insufficient capacity of the thin-walled web plate to be compressed along the path of compression (Fig. 4.39).

The diagonal shear buckles tend to be elongated in the tension direction, due to that the free width is larger there. This can initially seem a little strange, but if one accepts the fact that it is the loaded width that is the governing parameter with respect to buckling – and not the length – it is more easily understood. The buckle is quite simply forming itself in the less rigid zone.

The critical buckling stress with respect to pure shear loading (see Fig. 4.40), for a web panel (i.e. the free area, with the length a, between two vertical stiffeners), becomes (Eq. 4.22):

$$\tau_{cr} = k_\tau \cdot \frac{\pi^2 \cdot E}{12 \cdot (1 - v^2) \cdot \left(\frac{d}{t}\right)^2} \tag{4.22}$$

The buckling coefficient depends on the panel aspect ratio a/d (compare Fig. 2.13) (Eqs. 4.23, 4.24):

$$k_\tau = 5.34 + \frac{4}{\left(\frac{a}{d}\right)^2} \quad \left[\frac{a}{d} \geq 1\right] \tag{4.23}$$

$$k_\tau = 4 + \frac{5.34}{\left(\frac{a}{d}\right)^2} \quad \left[\frac{a}{d} \leq 1\right] \tag{4.24}$$

Just as we have learnt to understand by the truss analogy, the shear forces will cause one diagonal in tension, and one in compression (Fig. 4.41).

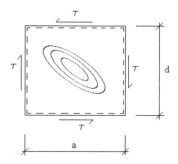

Figure 4.40 The buckling mode of a shear panel.

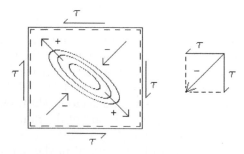

Figure 4.41 The action of a shear panel is more or less similar to that of a truss.

The shear deformation tends to transform the rectangular plate into a parallelogram, due to the shortening of the compression diagonal, and the elongation of the diagonal in tension (Fig. 4.42).

This shear deformation model will give us the chance to remind ourselves of the mode of action of the inner statically indeterminate truss. The diagonal that goes to the upper left corner, in the left-hand panel of the truss, is holding up the elongation of the panel – and the transverse diagonal, that goes to the lower left corner, i.e. the support, is holding up the compression. Should either one of these diagonals be damaged – e.g. the diagonal in compression (due to buckling for example) – would the truss (read: the panel) still be able to carry the loading (read: the diagonal in tension would now carry all the force alone). The inner statical system has gone from statically indeterminate to statically determinate (Fig. 4.43).

However, would also the diagonal in tension for some reason be damaged (e.g. through a hit damage), then the system will be a mechanism, and the entire truss would collapse (Fig. 4.44).

With reference to truss diagonals in tension or compression, we could here very shortly refer to some structures that we have discussed before. In the section regarding the Britannia Bridge it was mentioned about a patented truss bridge system (Pratt), where the diagonals were positioned in such a way that they always were in tension (as the bridge was loaded), and this meant that the bridge weight was kept to a minimum, as the diagonals not had to be designed for a possible buckling risk. And in the last section regarding concentrated loads, there was a discussion about the launching of a truss bridge in Mexico. Due to the reversed loading situation for the diagonals during erection (in relation to the action of the same for the final and simply supported bridge),

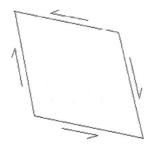

Figure 4.42 The resulting deformation of a shear panel.

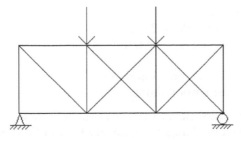

Figure 4.43 If one of the diagonals in an end panel is lost the truss is still functioning (given that the alternative load path – in this case the tension diagonal – is sufficiently strong).

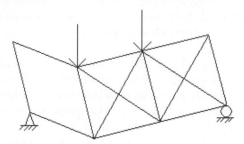

Figure 4.44 If both the diagonals in the end panel are lost then the truss becomes a mechanism and collapses.

the diagonals had to be braced, as they became subjected to compression during the launching, instead of being in tension as they would be for the final bridge (compare Fig. 4.23).

But let us resume to the discussion above for an inner statically indeterminate truss. The capacity to carry load in two alternative load-paths is exactly how the action in I-girders are assumed to be in the tension field method, with respect to shear loading. The maximum capacity in the ultimate loading state consists of two parts – first a part that is governed by the shear buckling strength (read: the diagonal in compression), and then a second (post-critical) part that is represented by the diagonal in tension.

The first contribution to the shear buckling resistance (and also the part that collapses first – in the sub-critical range) is given by the expression:

$$\Delta V = \tau_{cr} \cdot d \cdot t$$

To this capacity we have to add the post-critical reserve strength (read: the capacity of the tension field, going in the opposite direction of the compressions diagonal), and in order to find the maximum value of this second contribution, we study the possible band width g of the tension field, in a web panel having the length a, given a certain anchorage length s in the tension flange (index t) and in the compression flange (index c) (Fig. 4.45), (Eq. 4.25):

$$g = d \cdot \cos \varphi - (a - s_c - s_t) \cdot \sin \varphi \tag{4.25}$$

g width of the tension field
d depth of the web plate panel
φ inclination of the tension field ($\theta/2 \leq \varphi \leq \theta$)
θ slope of the panel diagonal (i.e. arc tan d/a)
a length of the panel
s_c anchorage length in the compression flange
s_t anchorage length in the tension flange

The inclination φ of the tension field varies between the values $\theta/2$ and θ. The minimum value is for the case when the flanges already are fully utilized in carrying a bending, and thus have no capacity left to, in addition, anchor a tension field (i.e. $s = 0$). The maximum value is for the case when the flanges are, to their full length (i.e. $s = a$), active

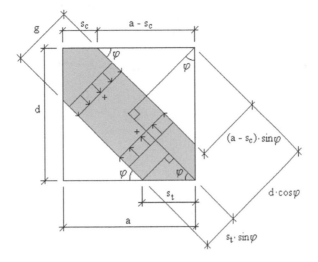

Figure 4.45 Possible band width, *g*, of a tension field in a shear panel.

in anchoring a tension field (read: complete tension field condition). The maximum value for the design shear buckling resistance in a given panel is found by iteration, in order to find the optimum inclination of the tension field between the limits. As an approximation, that is on the safe side, is $\varphi = \theta/1.5$. The tension field theory is only applicable for I-girders where the panel aspect ratios of the girder lie within the limits of $1 \leq a/d \leq 3$ – outside these limits the tension field theory is not valid. Besides that the tension field is anchored to the upper and lower flange, it is also anchored to the vertical stiffeners. These have then to be designed to carry this additional vertical reaction force they are subjected to (i.e. the vertical component of the tension field force). Just as a vertical stiffener under a concentrated load, the vertical stiffeners that are anchoring a tension filed have to be checked for the global buckling risk (i.e. out-of-plane buckling – relative to the web plate – in between the flanges). With respect to maximum possible anchorage length to the flanges (s_c and s_t), so is this length governed by a plastic hinge mechanism, where the plastic resistance moment of the flange is determining the possible anchorage length.

We study the part of the tension field that is anchored to the upper flange (Fig. 4.46), (Eqs. 4.26–4.28):

$$R = \sigma_{bb} \cdot s \cdot \sin \varphi \cdot t \cdot \sin \varphi \tag{4.26}$$

$$M_{pl} + M_{pl} = R \cdot \frac{s}{2} \tag{4.27}$$

$$\Rightarrow \quad s = \frac{2}{\sin \varphi} \cdot \sqrt{\frac{M_{pl}}{\sigma_{bb} \cdot t}} \tag{4.28}$$

The plastic hinge, at the distance *s* from the corner, has by definition a shear force equal to zero. The maximum moment in the "span region" is, according to the bending

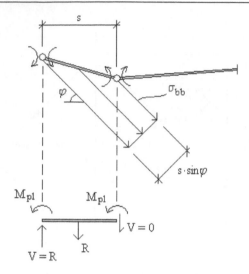

Figure 4.46 The capacity to anchor the tension field in the flange(s) is governed by the bending stiffness of the same.

theory, also found where the shear force $(\partial M/\partial x)$ is zero or changing sign. One way of looking at it is to accept the fact that all loads in the (truss) diagonal have to be attached to the (truss) joint.

s anchorage length in the flange
φ inclination of the tension field
M_{pl} plastic resistance moment of the flange $(W_{pl} \cdot f_y)$
W_{pl} plastic section modulus of the flange $(b_f \cdot t_f^2/4)$
b_f flange width
t_f flange thickness
σ_{bb} strength of the tension field (obtained from a yield criterion)
t thickness of the web plate

The plastic resistance moment of the flange has to be reduced allowing for a longitudinal force (from a bending moment or an externally applied axial force). The unreduced value above $(M_{pl} = W_{pl} \cdot f_y)$ is only used for the case of pure shear. The maximum capacity of the tension field (σ_{bb}) is given when full yielding over the band width g is reached. This strength was defined in the Eurocode (the first version of the prestandard) to (Eqs. 4.29, 4.30):

$$\sigma_{bb} = \sqrt{f_y^2 - 3 \cdot \tau_{cr}^2 + \psi^2} - \psi \qquad (4.29)$$

where:

$$\psi = 1,5 \cdot \tau_{cr} \cdot \sin 2\varphi \qquad (4.30)$$

In the final stages, when the compression diagonal has buckled, and the tension field has reached full yielding (and the flanges have collapsed in their plastic hinge

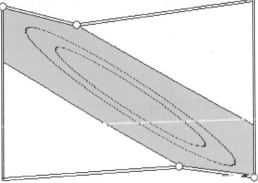

Figure 4.47 Collapse mode of a panel in shear.

mechanism), a compressed and considerably elongated buckle is formed along the tension field band (Fig. 4.47).

In order for the expression of the maximum possible anchorage length s in the flanges should be valid for an end panel – especially in the region where the tension field is anchored to the upper flange – it is required that the girder end is supplied with a double end post (Fig. 4.48).

In this way, a "continuity" is made for the upper flange over the inner stiffener (just as for the inner panels), which enables for the forming of a full plastic hinge. Without the extra stiffener at the end, the flexibility of the upper flange would reduce this capacity. Also the anchorage of the tension field to the vertical stiffener is gaining on this configuration, as the horizontal strength of the stiffener (over the support) is increasing. The rigid end post is acting as a short beam, resisting the horizontal component of the tension field band (Fig. 4.49).

When we finally have calculated the load-carrying capacity of the post-critical reserve effects (read: the capacity of the tension field), we add this to the shear buckling

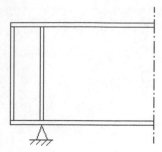

Figure 4.48 A double end post increases the shear buckling capacity.

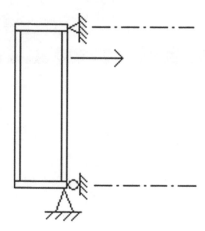

Figure 4.49 The double end post is acting as a short beam resisting the horizontal component of the inclined tension field.

capacity (read: the sub-critical strength), in order to receive the design shear buckling resistance in the ultimate limit state (having used all the capacity of both the compression diagonal and the tension diagonal) (Eq. 4.31):

$$V_{bb \cdot Rd} = \frac{\tau_{cr} \cdot d \cdot t + 0.9 \cdot (\sigma_{bb} \cdot g \cdot t \cdot \sin \varphi)}{\gamma_{M1}} \tag{4.31}$$

In the Eurocode (the first version of the prestandard), the maximum possible capacity of the tension field was reduced to 90% of the calculated maximum value. Two design methods were also given:

– The simple post-critical method. In order to calculate the design shear buckling resistance ($V_{ba \cdot Rd}$), an enhanced value for the shear strength is used, as a mean value of the shear buckling strength and the tension field strength taken together.
– The tension field method ($V_{bb \cdot Rd}$).

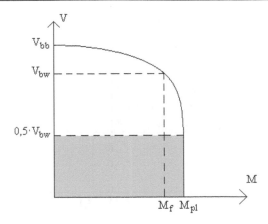

Figure 4.50 Combined shear/bending moment relationship; $V_{sd} \leq 0.5 \cdot V_{bw \cdot Rd}$.

For panels having a combined loading of shear and bending moment (which is normally the case), the following three criteria should be satisfied (depending on the magnitude of the design shear force):

$$V_{sd} \leq 0.5 \cdot V_{bw \cdot Rd}$$

The shear force does not exceed 50% of the *web only* shear buckling resistance, i.e. when the design shear buckling resistance is calculated having the tension field only anchored to the vertical stiffeners, not to the flanges (i.e. $s_c = s_t = 0$). The inclination of the tension field is $\theta/2$. The design bending moment resistance needs not to be reduced to allow for the shear force (Fig. 4.50).

$$0.5 \cdot V_{bw \cdot Rd} \leq V_{sd} \leq V_{bw \cdot Rd}$$

The shear force exceeds 50% of the web only shear buckling resistance, but not 100% of the same. When the bending moment exceeds the *flange only* bending moment resistance $M_{f \cdot Rd}$ – i.e. the bending moment capacity of the girder, consisting of the flanges only (neglecting the contribution from the web) – the following criterion should also be satisfied (Eq. 4.32):

$$M_{sd} \leq M_{f \cdot Rd} + (M_{pl \cdot Rd} - M_{f \cdot Rd}) \cdot \left[1 - \left(\frac{2 \cdot V_{sd}}{V_{bw \cdot Rd}} - 1 \right)^2 \right] \tag{4.32}$$

where $M_{pl \cdot Rd}$ is the design plastic resistance moment of the gross section, i.e. without considering any local buckling risk giving an effective cross-section (Fig. 4.51).

$$V_{sd} \geq V_{bw \cdot Rd}$$

The design shear buckling resistance is calculated with respect to the complete tension field method (i.e. $V_{bb \cdot Rd}$), where the combined effect of shear and moment is considered in the maximum possible anchorage length in the flanges (Fig. 4.52).

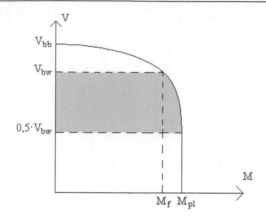

Figure 4.51 Combined shear/bending moment relationship; $0.5 \cdot V_{bw \cdot Rd} \leq V_{sd} \leq V_{bw \cdot Rd}$.

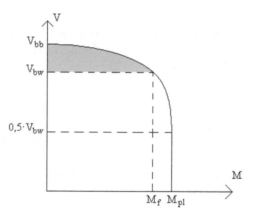

Figure 4.52 Combined shear/bending moment relationship; $V_{sd} \geq V_{bw \cdot Rd}$.

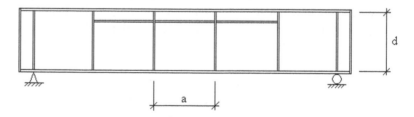

Figure 4.53 Longitudinal stiffeners increase the shear buckling strength.

We have until now only discussed shear panels without longitudinal stiffeners, but often have such stiffeners to be considered. In simply supported bridge girder spans, the mid-span region are normally having horizontal (longitudinal) stiffeners in the upper part where the girder is subjected to compression from the bending moment (compare section 4.2), and these stiffeners are having a positive effect on the shear buckling strength (Fig. 4.53).

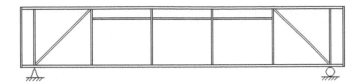

Figure 4.54 Diagonal stiffeners – following the compression flow in end panels – are especially efficient.

For these longitudinally stiffened panels, an enhanced value of the shear buckling coefficient can be calculated (Eqs. 4.33, 4.34):

$$k_\tau = 5.34 + \frac{4}{\left(\frac{a}{d}\right)^2} + \frac{2.1}{t} \cdot \sqrt[3]{\frac{I_{st}}{d}} \quad \left[\frac{a}{d} \ge 1\right] \tag{4.33}$$

$$k_\tau = 4 + \frac{5.34}{\left(\frac{a}{d}\right)^2} + \frac{2.1}{t} \cdot \sqrt[3]{\frac{I_{st}}{d}} \quad \left[\frac{a}{d} \le 1\right] \tag{4.34}$$

where I_{st} is the second moment of area of the longitudinal stiffener with respect to bending (buckling) in the out-of-plane direction relative to the web plate.

With respect to the end panels, which are subjected to more or less pure shear – and where longitudinal stiffeners are not needed due to low bending moments – one could consider diagonal stiffeners (just as Jourawski did suggest for the Britannia Bridge once). The optimal position of such a stiffener would be in the compression diagonal, such that this stiffener – and the longitudinal stiffeners – is more or less following the compression flow (Fig. 4.54).

A stiffener in the tension diagonal would not be as efficient, even if it is close at hand to believe so as the elongated buckle is orientated along this diagonal – the buckling would instead occur on either side of such a stiffener. The buckling coefficient for a square web panel (i.e. with $a/d = 1$), with or without diagonal stiffener, is:

9.34 without a stiffener
11.4 a diagonal stiffener in the tension diagonal
30 a diagonal stiffener in the compression diagonal

As to determine the shear force capacity of slender webs, the first version of the pre-standard (ENV) contained the full use of the tension field method (as it also has been presented here up to this point), however, for design purposes it has, most evidently, shown a little bit too complicated. The Eurocode has therefore been adapted in order to meet these needs. The expressions to use, and which are also to be presented in the following, simplifies the design process a great deal, however, the shear buckling phenomenon is somewhat hidden, which is unfortunate.

The design resistance, of a slender web taking shear buckling into account, $V_{b,Rd}$, according to the current version of the Eurocode (EN 1993-1-5: 2006) (Eq. 4.35):

$$V_{b,Rd} = V_{bw,Rd} + V_{bf,Rd} \le \frac{\eta \cdot f_{yw} \cdot h_w \cdot t}{\sqrt{3} \cdot \gamma_{M1}} \tag{4.35}$$

Table 4.1 The shear buckling factor, χ_w, as a function of the slenderness parameter.

	Rigid end post	Non-rigid end post
$\bar{\lambda}_w < 0.83/\eta$	η	η
$0.83/\eta \leq \bar{\lambda}_w < 1.08$	$0.83/\bar{\lambda}_w$	$0.83/\bar{\lambda}_w$
$\bar{\lambda}_w \geq 1.08$	$1.37/(0.7 + \bar{\lambda}_w)$	$0.83/\bar{\lambda}_w$

$V_{bw,Rd}$ contribution from the web
$V_{bf,Rd}$ contribution from the flanges
η 1.20 for steel grades $\leq$ S460
         1.0 for steel grades $>$ S460
f_{yw} yield strength of the web
h_w depth of the web (earlier denoted d, could also be denoted b_w)
t thickness of the web
γ_{M1} partial factor (instability)

The contribution from the web is given by (Eq. 4.36):

$$V_{bw,Rd} = \frac{\chi_w \cdot f_{yw} \cdot h_w \cdot t}{\sqrt{3} \cdot \gamma_{M1}} \tag{4.36}$$

The shear buckling factor, χ_w, is governed by a slenderness parameter (Eq. 4.37):

$$\bar{\lambda}_w = 0.76 \cdot \sqrt{\frac{f_{yw}}{\tau_{cr}}} \tag{4.37}$$

Besides being governed by the critical shear buckling stress, τ_{cr} (a lowered critical buckling stress value increases the slenderness parameter – the opposite for the yield stress), the shear buckling factor is also a function of the stiffeners surrounding the shear panel. For example do rigid end posts at the supports increase the capacity of the end panel to carry shear (read: increase the coefficient χ_w), at least for higher slenderness values (Table 4.1).

The contribution from the flanges (given that the flanges are not fully utilized in resisting the bending moment alone, i.e. $M_{Ed} < M_{f \cdot Rd}$, where $M_{f \cdot Rd}$ is the flange only moment of resistance) (Eq. 4.38):

$$V_{bf,Rd} = \frac{b_f \cdot t_f^2 \cdot f_{yf}}{c \cdot \gamma_{M1}} \cdot \left[1 - \left(\frac{M_{Ed}}{M_{f,Rd}} \right)^2 \right] \tag{4.38}$$

where:
b_f, t_f width and thickness of the flange which provides the least axial resistance (the width of the flange should also be limited to $15 \cdot \varepsilon \cdot t_f$ on each side of the web). $\epsilon = (235/f_y)^{0.5}$
f_{yf} yield strength of the flange

c anchorage length in the flanges of the tension field
(compare pages 94–97)

γ_{M1} partial factor (instability)

M_{Ed} design bending moment

$M_{f,Rd}$ *flange only* bending moment of resistance $M_{f,Rd} = M_{f,k}/\gamma_{M0}$.
If there is an axial force present, the flange only bending moment of
resistance should be reduced.

γ_{M0} partial factor (resistance of cross-sections)

The anchorage length, c, is equal to (Eq. 4.39):

$$c = a \cdot \left(0.25 + \frac{1.6 \cdot b_f \cdot t_f^2 \cdot f_{yf}}{t \cdot h_w^2 \cdot f_{yw}} \right) \tag{4.39}$$

where a is the distance between the vertical stiffeners.

The effect from longitudinal stiffeners (in between the vertical stiffeners) is an
increase of the buckling coefficient k_τ (Eqs. 4.40–4.42):

$$k_\tau = 5.34 + \frac{4}{\left(\frac{a}{h_w}\right)^2} + k_{\tau st} \qquad \left[\frac{a}{h_w} \geq 1 \right] \tag{4.40}$$

$$k_\tau = 4 + \frac{5.34}{\left(\frac{a}{h_w}\right)^2} + k_{\tau st} \qquad \left[\frac{a}{h_w} \leq 1 \right] \tag{4.41}$$

where:

$$k_{\tau st} = 9 \cdot \left(\frac{h_w}{a}\right)^2 \cdot \left(\frac{I_{sl}}{t^3 \cdot h_w}\right)^{3/4} \qquad \left[\geq \frac{2.1}{t} \cdot \left(\frac{I_{sl}}{h_w}\right)^{1/3} \right] \tag{4.42}$$

$I_{sl} =$ the second moment of area (in the out-of-plane direction relative to the web plate)
of the longitudinal stiffener.
(compare page 101)

The critical buckling stress – for a panel having a longitudinal stiffener – should not
be taken less than for the largest subpanel (i.e. the largest part of the web above or
below the longitudinal stiffener).

If the design shear force, V_{Ed}, exceeds 50% of the web only shear buckling resistance, $V_{bw,Rd}$, and the design bending moment, M_{Ed}, exceeds the flange only bending
moment of resistance, $M_{f,Rd}$, then the following interaction criterion should be satisfied
(compare pages 99–100) (Eqs. 4.43, 4.44):

$$\bar{\eta}_1 + \left(1 - \frac{M_{f,Rd}}{M_{pl,Rd}} \right) \cdot (2\bar{\eta}_3 - 1)^2 \leq 1.0 \tag{4.43}$$

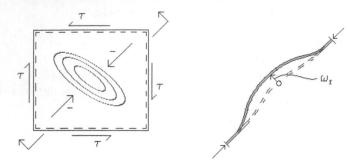

Figure 4.55 An already present out-of-plane imperfection will "breathe" back and forth as the web plate is subjected to repeated shear loading.

where:

$$\bar{\eta}_1 = M_{Ed}/M_{pl,Rd} \qquad\qquad (4.44)$$

$M_{pl,Rd}$ is the design plastic resistance of the cross-section of the effective area of the flanges and the fully effective web irrespective of its section class;

$$M_{pl,Rd} = W_{pl} \cdot f_y/\gamma_{M0} \qquad\qquad (4.45)$$

$$\bar{\eta}_3 = V_{Ed}/V_{bw,Rd} \qquad\qquad (4.46)$$

The interaction criterion above does not have to be verified in sections located at a distance less than $h_w/2$ from a support with vertical stiffeners.

Web breathing

As a bridge over from the shear buckling risk (in the compression diagonal) to a special kind of problem in the tension field diagonal, the phenomenon of web breathing is worth mentioning. As the approach in the modern codes, with respect to maximum load-carrying capacity of thin-walled plated structures (taking post-critical reserve effects into account, have become more and more up-to-date with the full knowledge regarding the buckling behaviour, the plates have also become thinner. Due to a much greater knowledge regarding the static behaviour of structures with respect to different local instability phenomenon, which is reflected in the design, there is on the other hand a possible risk that other types of problems do occur. The local instability problem is treated in a correct and optimal way, however, the fatigue problem due to the tension loading tends to be more complex and difficult to handle. We consider an unstiffened web panel having initial imperfections, and being subjected to repeated loading (e.g. from the passing of trains on a railway bridge). An already existing small out-of-plane curvature of the web plate will bend even more as the loading is introduced (Fig. 4.55).

The shear loading coming from the train will be added to the shear loading due to self-weight of the bridge itself, which give us the shear stress range τ_r (i.e. the difference between the two). It could very well be a matter of small loads in relation to the maximum load-carrying capacity (in the diagram below this particular action is enlarged for the sake of understanding) (Fig. 4.56).

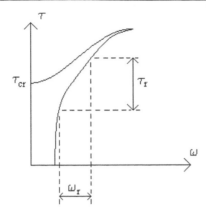

Figure 4.56 Shear loads well below the maximum load-carrying capacity could produce large out-of-plane deflections back and forth.

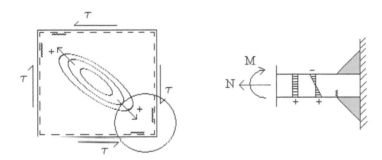

Figure 4.57 Cracks are eventually formed where the web plate is fixed (and where the combined effect of tension and local bending is the highest).

This back and forth out-of-plane deflection of the web plate can – if the number of loading cycles is large enough – cause fatigue cracking, even if it is, for the naked eye, a matter of small "web breathing movements". The fatigue cracks that have been observed in shear panels (i.e. panels with dominating shear), have been located where the tension field is anchored to the flange and the vertical stiffener (Fig. 4.57).

The cracks are initiated at the weld toe – on the opposite side of the web plate relative to the direction of the deflection – due to the combined effect of weld defects, residual stresses, and fluctuating bending and normal stresses. The cracks propagate through the thickness of the web plate and also along the weld itself, and if they are left unattended the load-carrying capacity could be reduced, as the anchorage of the tension field becomes decreased. If the cracks are growing in contact with each other – across the corner of the panel – the capacity of the entire tension field is gone.

Similar cracks have also been discovered in mid-span region panels having dominating bending moment, however, in this case caused by back and forth deflection of normal stress buckles. These types of cracks have – due to the nature of the loading – only been observed in the intersection between the web plate and the compression flange, because of the position of the buckles.

Chapter 5

Shell buckling

5.1 Introduction

In structural engineering there is one particular cross-section that definitely is more optimal than others, and that is the shell. A shell, that either could be of single curvature or double curvature shape, is the most optimal form when it comes to carrying compression forces in its membrane, and it is the shape itself that stabilizes against buckling. There are two principal ways a single curvature shell can carry load, and that is either in the longitudinal direction (for axial loading), or in the transverse direction (for an evenly distributed, externally applied pressure – alternatively, an internally applied suction) (Fig. 5.1).

If the loading of a shell is coming from an internally applied pressure, the membrane becomes subjected to tension, instead of compression. An example of such a pressure vessel – where the tension force in the membrane was accidentally reversed to compression, and because of this, buckling of the shell did occur (as the same was not designed to carry such loading) – was the butane gas train carriages that "imploded" 1976 in France. Twenty carriages had been emptied on their content (the butane gas) in the south of France, near Marseille, at a temperature of +18°C, and were thereafter transported to the German border in the north. At the time of the arrival, the temperature had dropped to −24°C, an extremely low temperature for French weather conditions it must be admitted, however, still quite possible as it was in January. Even though that the carriages had been emptied on their content, they still contained some

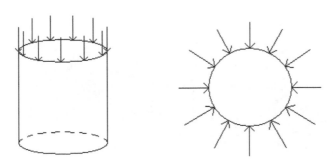

Figure 5.1 A single curvature shell can carry load either in the longitudinal (axial) or transverse direction.

remaining gas, which did condensate as it was cooled off. This considerable cooling, and the subsequent condensation, led to a sub-pressure of the shell containers, which they had not been designed for. The compression forces in the membrane became too much for the shells to carry, and they just collapsed (inwards) (Fig. 5.2).

This surprising incident did show the following; first of all that a shell must be designed to carry the compressive forces it is (by extraordinary loading or not) going to be subjected to, and secondly, that buckling of a shell is a very sudden and collapse-like phenomenon.

We shall in the following concentrate on cylinders (i.e. single curvature shells), where the dominating load is axial compression, and such profiles are common in bridge engineering constructions because of their optimal characteristics. If we consider a compression diagonal in a truss bridge, then a hollow circular section has the advantage that the (global) buckling resistance is equally large in all directions. In addition, the wind resistance is low, the surface area is minimized (reduced maintenance and painting costs, and the aesthetics is higher relative other standard rolled profiles. However, a certain consideration has to be made in the design of cylinder profiles, and that is to put extra care in treating the local buckling risk.

Historically, this local buckling risk has been treated with great success, and there are many impressive bridges using tubular sections. Royal Albert Bridge, over the river Tamar at Saltash in the south-west of England, was built by Isambard Kingdom Brunel in 1859. This epoch-making bridge has two centre spans of 140 meters each, and these spans have an upper arch-shaped compression chord of a closed tube section, that interacts with a lower "chain link bar" in tension. The simplified static behaviour that was discussed in section 4.4 with respect to I-girders is used by its pure form in this bridge. The effect of the combined local and global buckling risk is really shown

Figure 5.2 The unexpected implosion in 1976 of some train carriages in France.

on the upper chord dimension, which has become extremely huge in comparison to the lower chord (where no instability considerations had to be made as it is subjected to tension only) (Fig. 5.3).

Another, and equally as epoch-making tubular bridge, is the railway bridge over the Firth of Forth, just north of Edinburgh in Scotland (Fig. 5.4).

Figure 5.3 The Royal Albert Bridge at Saltash.

Figure 5.4 The Forth Bridge.

This cantilevering girder bridge that was built in 1890 became the longest bridge in the world. The centre spans consist of two huge cantilevers, each 207 meters long, supporting a suspended mid-section of 107 meter, making up a total span length of 521 meters. At the time when the bridge was to be constructed, the designers had to follow the strict regulations that stipulated that the maximum stress was not to exceed one fourth of the strength of the steel material, irrespective of the stress being in tension or compression. The bridge was given its huge dimensions not only due to this demand, but also because of the Tay Bridge disaster some ten years earlier. The railway bridge over the Firth of Tay did collapse during a gale in late December 1879, and this incident hade a great impact on the stability requirements and the load-carrying capacity of bridges to come.

5.2 Theory

The elastic critical buckling stress of an ideal and perfect cylinder, axially loaded with an evenly distributed compression load, is equal to (Fig. 5.5) (Eq. 5.1):

$$\sigma_{cr} = \frac{1}{\sqrt{3 \cdot (1 - v^2)}} \cdot E \cdot \frac{t}{r} = \underline{0.605 \cdot E \cdot \frac{t}{r}} \tag{5.1}$$

(with $v = 0.3$ for steel)

As a comparison between the load-carrying capacity of a cylinder, and that of a plate and a strut, we consider a rectangular aluminium plate (having a width of 204 mm, a height of 150 mm, and a thickness of 0.18 mm) that is subjected to an evenly distributed axial load (Fig. 5.6).

If only the loaded edges are supported, we regard the plate as a *strut*, for which we can calculate the critical (global) buckling load (Eqs. 5.2, 5.3):

$$\sigma_{cr} = \frac{\pi^2 \cdot E}{12 \cdot (1 - v^2) \cdot \left(\frac{150}{0.18}\right)^2} = \underline{0.093 \, \text{MPa}} \tag{5.2}$$

$$\Rightarrow \quad P_{cr} = 0.093 \cdot 204 \cdot 0.18 = \underline{3.4 \, \text{N}} \tag{5.3}$$

(with $E = 70,000 \, \text{MPa}$ and $v = 0.33$ for aluminium)

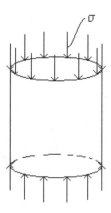

Figure 5.5 An axially loaded cylinder.

If also the unloaded edges are supported, we have, by definition, a *plate*, which has a critical buckling stress equal to (Eqs. 5.4–5.6):

$$\sigma_{cr} = 4.39 \cdot \frac{\pi^2 \cdot E}{12 \cdot (1 - v^2) \cdot \left(\frac{204}{0.18}\right)^2} = \underline{0.221\,\text{MPa}} \tag{5.4}$$

$$\text{with } k = \left(\frac{204}{150} + \frac{150}{204}\right)^2 = \underline{4.39} \tag{5.5}$$

$$\Rightarrow \quad P_{cr} = 0.221 \cdot 204 \cdot 0.18 = \underline{8.1\,\text{N}} \tag{5.6}$$

However, if we fold the plate together into a closed cylinder, having a diameter of 65 mm ($204/\pi$) – similar in shape to that of a beer can – the critical buckling stress will increase dramatically (and so will the critical load) (Fig. 5.7) (Eqs. 5.7, 5.8).

$$\sigma_{cr} = \frac{1}{\sqrt{3 \cdot (1 - v^2)}} \cdot E \cdot \frac{t}{r} = \underline{237.1\,\text{MPa}} \tag{5.7}$$

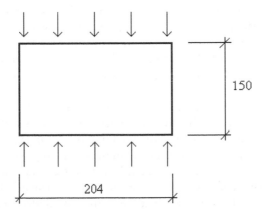

Figure 5.6 An axially loaded simple plate.

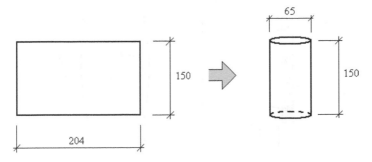

Figure 5.7 The transformation of the plate into a closed cylinder.

$$\Rightarrow \quad P_{cr} = 237.1 \cdot 204 \cdot 0.18 = \underline{8707\,\text{N}} \qquad\qquad (5.8)$$

The theoretical load-carrying capacity (with respect to buckling) has, as a consequence of the transformation of the aluminum foil from a plate (and a strut) to a cylinder, increased from less than a kilogram to almost 900 kg (!). And if we had made two cylinders of the plate (with a diameter of 32.5 mm each), the capacity would have been doubled (as the critical buckling stress increases linearly to the decrease in slenderness r/t). There are, however, two considerations to make for the cylindrical shell. First of all the fact that the critical buckling stress markedly (to say the least) exceeds the yield strength of about 30 MPa, and secondly that the classical buckling theory also markedly overestimates the load-carrying capacity of true structures, as the effect of initial imperfections have a large reducing influence, especially for cylinders having a large slenderness. We will come back to the actual design of true cylinders further on in this section, however, first we will concentrate on the actual behaviour of the cylinder, trying to disentangle why there is such a large difference in results between the load-carrying capacity of a cylinder in relation to that of a strut or a plate.

If a plate did show its post-critical reserve behaviour after buckling had occurred, the similar stabilizing forces has already been used before buckling of axially loaded cylinder shells. For plates, transverse membrane forces in tension did stabilize the buckle and made it possible for additional loading in the post-critical range. However, for a cylindrical shell, the membrane is not only stabilized by transverse tension forces, but also by transverse compression forces. It is the circular form of the cylinder that enables this optimal ability of alternating stabilization of the membrane, and it is also the explanation why a shell is able to carry such extreme loading, at least according to the theory.

If we consider a strip in the longitudinal direction of the cylinder, it is consequently stabilized by forces in the circumference direction that either are in tension or compression depending on the possible buckling direction (Fig. 5.8).

As the critical buckling stress is reached, these stabilizing forces in the transverse direction are already used to the full, which has the effect that there is no additional reserve strength to consider (as is the case for plates), instead the ultimate load is followed by a sudden and instant collapse (note that the deflection on the horizontal

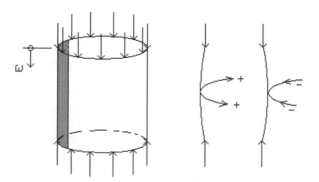

Figure 5.8 Stabilizing membrane forces in the transverse circumferential direction.

axis is in the longitudinal direction of the cylinder, not in the out-of-plane direction as was the case for the load/deformation-curve of plates) (Fig. 5.9).

According to the classical theory, there is an additional loading capacity (after collapse) due to the symmetrical buckling pattern, but this "sub-critical branch" exists only in theory, and possibly also for deformation induced load tests of more or less perfect shells. For real shells, having imperfections, the load/deformation-curve shows a completely different behaviour (Fig. 5.10).

True and imperfect cylinder shells do buckle at a markedly lower level than what is predicted by the classical theory for ideal conditions. Already since the classical theory was introduced in the beginning of the last century, much lower tests results were observed in comparison to what had been predicted, and it was first during the

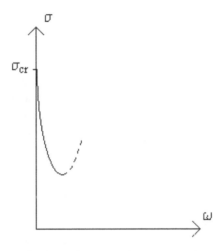

Figure 5.9 Stress/deformation relationship for an axially loaded perfect (i.e. ideal) cylinder.

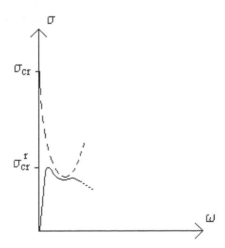

Figure 5.10 The stress/deformation relationship for an imperfect cylinder.

1950s that researchers could explain the reason for this deviation. Besides initial imperfections in the curved surface of the shell, also residual stresses and uneven loading
distribution do explain the difference in results. The theory had to be revised in accordance to these effects, and a reduced value of the critical buckling stress, from a large
number of load tests on cylinders having varying slenderness ratios, was established.
It was found, besides as expected, that the true critical buckling stress was markedly
lower than what the classical theory did predict, also that the slenderness did influence largely. As the slenderness did increase, the observed buckling stress decreased in
relation to the calculated value (Fig. 5.11).

The following experimentally determined relationship, between the classical buckling theory and the observed, was established (Eqs. 5.9, 5.10):

$$\sigma_{cr}^r = \alpha \cdot \eta \cdot \sigma_{cr} \tag{5.9}$$

$$\eta = \frac{0.83}{\sqrt{1 + 0.01 \cdot \frac{r}{t}}} \tag{5.10}$$

where α is considering the production method and tolerance level, and η is an experimentally set "knock-down-factor" with respect to the slenderness ratio of the shell.

That the observed value of the critical buckling stress is decreasing more relative to
the classical theory, for high slenderness ratios, depends upon the fact that the inner
stabilizing membrane forces become more sensitive for imperfections as the slenderness is increasing (read: as the curvature is decreasing, the circumference forces become
more easy to disturb).

A disturbed geometry is affecting the load-carrying capacity in different ways. Outwards going initial imperfections in the shell have a tendency to be held back as
membrane forces in tension are activated (Fig. 5.12).

Inwards going imperfections, on the other hand, create not only a disturbance that
gives local bending of the shell, but is also being "pushed on" by the membrane stresses,
which now are in compression (Fig. 5.13).

Initial imperfections – especially the inwards going – do function as small local disturbances, which reduce the ability to stabilize the shell through membrane (compression)
forces. Around a defect there is also a redistribution, which increases the loading in
the surrounding areas (Fig. 5.14).

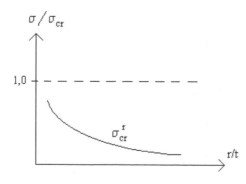

Figure 5.11 The true critical buckling stress is highly dependent on the slenderness ratio, r/t.

This lowering effect on the load-carrying capacity, that especially was the result of the inwards going imperfections, also explains why shells are so sensitive for transverse loading (and the greater slenderness the worse). If an out-of-plane transverse load is applied at the same time as the shell is subjected to a large axial load, there is a risk that the stabilizing membrane forces are completely knocked out, and as a consequence a collapse will occur instantly. The shell will go from one equilibrium state to another (and also being a heavily deformed state) (Fig. 5.15).

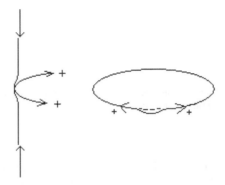

Figure 5.12 Outwards going imperfections tend to be held back by the membrane forces (being in tension).

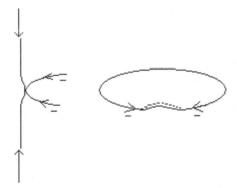

Figure 5.13 Inwards going imperfections tend to be "pushed on" by the membrane forces (being in compression).

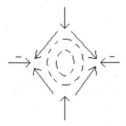

Figure 5.14 The load dispersion around a local defect increases the compression in this zone.

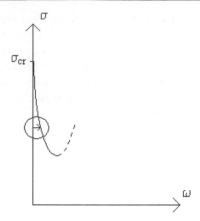

Figure 5.15 The sudden "jump" from equilibrium to a collapse by the introduction of a small additional disturbance of the stabilizing membrane forces.

This sensitivity against additional disturbances coming from concentrated loads is a well-known behaviour of thin and slender shells, a phenomenon that also the designers are fully aware of. If a shell is optimized with respect to an evenly distributed load, the thickness becomes normally very thin, and as a consequence the capacity against additional transverse loading becomes heavily reduced. When the Stockholm Globe Arena was to be constructed in 1986, there was a problem regarding the platform for sound and lighting. Would this platform be suspended directly to shell, there would be a risk for the introduction of large concentrated forces (due to unevenly distribution of the load on the platform). The problem was solved in such a way that the suspension cables were attached to the top of a vertical steel column (being only supported in the transverse direction at the top), which in its turn transferred the total load down to a new set of cables, which carried the load (now evenly distributed) to the connection points in the shell (Fig. 5.16).

If we go back to the "can" that was analyzed in the beginning of this section, we can calculate the design load-carrying capacity with the help of the "Shell Handbook", i.e. its true resistance. The cylinder was 150 mm high, had a diameter of 65 mm, and a thickness of the wall of 0.18 mm, and it was made of aluminium plate, having a yield strength of 30 MPa, a modulus of elasticity of 70,000 MPa, and a Poisson's ratio of 0.33 (Eq. 5.11) (Fig. 5.17).

$$\sigma_{cr} = \frac{1}{\sqrt{3 \cdot (1 - v^2)}} \cdot E \cdot \frac{t}{r} = \underline{\underline{237.1 \text{ MPa}}} \tag{5.11}$$

(as before)

As the length in relation to the diameter is relatively small there is no risk of global (Euler) buckling, instead we concentrate on the risk of local shell buckling. The capacity will be calculated assuming safety class 1 according to the Swedish code, i.e. $\gamma_n = 1.0$. The reduction factor α, which took into account the production method and tolerance level, is determined by the help of Table 5.1.

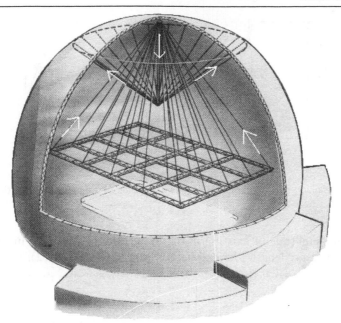

Drawing: Ingemar Franzén

Figure 5.16 The ingenious rigging of the sound and lighting platform of the Stockholm Globe Arena.

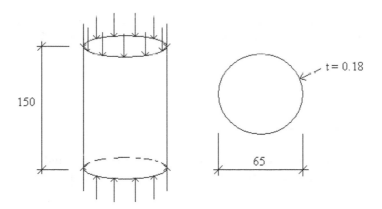

Figure 5.17 A 65 mm wide and 0.18 mm thick axially loaded aluminium cylinder.

Table 5.1 The reduction factor, α.

Production method	Tolerance class	ω_r	α
Hot formed/Stress relieved	1	$\leq l_r/100$	1
	2	$\leq l_r/50$	0.7
Cold formed/Welded	1	$\leq l_r/100$	0.9
	2	$\leq l_r/50$	0.6

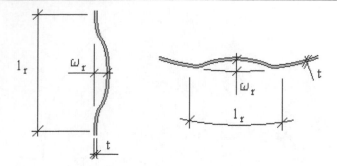

Figure 5.18 The definition of how to determine the magnitude of the initial imperfections.

where ω_r is the amplitude of existing initial imperfections (buckles) in the shell, and where l_r is the measure length (defined below) (Fig. 5.18) (Eq. 5.12).

$$l_r = 4 \cdot \sqrt{r \cdot t} \qquad (5.12)$$

We start with the assumption of tolerance class 2 $(\omega_r \leq l_r/50)$ of our cold-formed aluminium cylinder. This gives the reduction factor:

$$\alpha = 0.6$$

In our specific case, this tolerance class would mean a maximum amplitude of existing initial imperfections that is comparable to the wall thickness (Eqs. 5.13, 5.14):

$$l_r = 4 \cdot \sqrt{32.5 \cdot 0.18} = \underline{9.7\,\text{mm}} \qquad (5.13)$$

$$\Rightarrow \quad \omega_r \leq \frac{9.7}{50} = \underline{0.19\,\text{mm}} \qquad (5.14)$$

Such a tight tolerance is probably not to be expected for beer cans and such, so instead we assume a maximum initial imperfection equal to twice the wall thickness (Eq. 5.15):

$$\omega_r \leq 2 \cdot t = \underline{0.36\,\text{mm}} \qquad (5.15)$$

and as this study of ours is an analysis, and not a design, we are able to calculate a reduction factor α', despite the fact that we are above the regulated tolerance values. The expression below will give us the new value of the reduction factor (Eq. 5.16):

$$\alpha' = \alpha - 15 \cdot \left(\frac{\omega_r}{l_r} - \frac{1}{50} \right)$$

$$= 0.6 - 15 \cdot \left(\frac{0.36}{9.7} - \frac{1}{50} \right) = \underline{0.343} \qquad (5.16)$$

This new expression given above is valid for initial imperfections up to the value of (Eq. 5.17):

$$\omega_r \leq \frac{l_r}{25} = \frac{9.7}{25} = \underline{0.39\,\text{mm}} \tag{5.17}$$

(>0.36 mm OK!)

The knock-down factor with respect to the slenderness ratio (Eq. 5.18):

$$\eta = \frac{0.83}{\sqrt{1 + 0.01 \cdot \frac{32.5}{0.18}}} = \underline{0.496} \tag{5.18}$$

The reduced value of the critical buckling stress can now be calculated (Eq. 5.19):

$$\sigma_{cr}^r = 0.343 \cdot 0.496 \cdot 237.1 = \underline{40.3\,\text{MPa}} \tag{5.19}$$

Depending on the level of this reduced value of the critical buckling stress, in relation to the yield strength of the material, the design strength resistance is calculated according to (Eqs. 5.20–5.23):

$$\sigma_{cr}^r \leq \frac{f_y}{3} \quad \Rightarrow \quad f_{rd} = 0.75 \cdot \sigma_{cr}^r \tag{5.20}$$

$$\sigma_{cr}^r \geq \frac{f_y}{3} \quad \Rightarrow \quad f_{rd} = \omega_s \cdot f_y \tag{5.21}$$

$$\text{where} \quad \omega_s = \frac{0.96}{0.8 + \lambda_s^2} \quad (\leq 1.0) \tag{5.22}$$

$$\text{and} \quad \lambda_s = \sqrt{\frac{f_y}{\sigma_{cr}^r}} \tag{5.23}$$

In our case $\sigma_{cr}^r > f_y/3$ (even above f_y), which gives (Eqs. 5.24, 5.25):

$$\lambda_s = \sqrt{\frac{30}{40.3}} = \underline{0.863} \tag{5.24}$$

$$\Rightarrow \quad \omega_s = \frac{0.96}{0.8 + 0.863^2} = \underline{0.621} \tag{5.25}$$

The design strength resistance (Eq. 5.26):

$$f_{rd} = 0.621 \cdot 30 = \underline{18.6\,\text{MPa}} \tag{5.26}$$

and the maximum load-carrying capacity (Eq. 5.27):

$$P_{sd} = 18.6 \cdot 204 \cdot 0.18 = \underline{683\,\text{N}} \tag{5.27}$$

This loading capacity of approximately 70 kg is about what can be expected of a beer can or similar, and it is relatively easy to check the result by yourself at home to see that it is correct. If someone is asked to assist during the test, you could also take the opportunity to ascertain the sensibility against transverse loading by lightly tapping on the side of the can as you are standing on the same – however, be prepared for the sudden collapse (and be using an empty can!).

The discussion so far has been concentrated on unstiffened cylinder shells, but there are a number of ways of how to improve the capacity by the use of different stiffeners. We have seen that unstiffened cylinder shells possess an inherent capacity of some magnitude to stiffen itself by the use of membrane forces, but we have also seen that these stabilizing forces are easily disturbed and knocked out, especially for shells having thin and slender walls. Ring stiffeners are primarily used for cylinders subjected to inner sub-pressure (or externally applied over-pressure). Such stiffeners have, however, small effect on cylinders subjected to axial loading, where instead longitudinal stiffeners should be chosen. If such stiffeners are chosen of a large dimension, there is a possibility to use post-critical reserve effects, just as for buckling of plane plates. Buckling in between the stiffeners does not mean the collapse of the shell, as additional loading can be accepted by yielding of the stiffer zones alongside of the stiffeners. The capacity of the longitudinal stiffeners can be increased by the use of additional ring stiffeners, which reduce the buckling length of the former – just as the function of longitudinal and transverse stiffeners at I-girders or box-girders. Older bridges having tubes (as the old riveted bridges at Saltash and Edinburgh that have been mentioned earlier) used both longitudinal and transverse (ring) stiffeners. However, in these old structures it was natural to insert both types of stiffeners, as the tubes were put together by smaller (and curved) plates. More modern truss bridges, having circular hollow sections, are without both longitudinal and ring stiffeners, as these bars are normally of a small dimension of a rolled standard profile. However, in modern steel arch bridges, having a cylinder profile, the section is built up of smaller segments, and then it is common practice to insert ring stiffeners at the joints, but not longitudinal stiffeners in between.

A modern steel arch bridge that definitely also would have been in need of longitudinal stiffeners (and not only ring stiffeners) was the Almö Bridge, also called the Old Tjörn Bridge. At the early stages of the design of the bridge in the 1950s, the suspension bridge alternative was rejected, as it was considered not to interact with the landscape of the province of Bohuslän, and also to disturb and endanger the air traffic in the area by the pylons. Instead the structure was turned upside down (reversing the stress flow from tension to compression), and a huge arch bridge was chosen, spanning over the entire channel. The span was impressive 278 meters, however, despite the rise of the arch being 41 meter, the free sailing width was quite limited (Fig. 5.19).

On the 18th of January 1980, the inevitable thing, that more or less had been a high probability during 20 years in service, occurred (read: that a ship easily can come off course in heavy fog). The cargo-ship Star Clipper lost its route through the channel of the Almö Sound, and hit the side of the arch by a gantry crane up at the bridge of the ship. It is hard to judge whether this applied load was big or small, but from the shell buckling theory we have learnt that shells are very sensitive to transverse forces, especially if they are – at the same time – subjected to an axial load

Jättetankern "THORSHAMMER" på 228.250 ton passerar Almöbron

Figure 5.19 "The giant tanker 'THORSHAMMER' of 228,250 tons passing the Almö Bridge".

Figure 5.20 As the cargo ship Star Clipper came off course in the fog, the tube arches were hit by the gantry crane of the ship, and the bridge collapsed.

of some magnitude. The arches of the Almö Bridge did, nevertheless, not withstand this additional transverse loading, and did collapse and fell into the sound (Fig. 5.20).

After such a major incident as this, the understanding definitely becomes greater with respect to the demands of the Admiralty concerning the free sailing width and height of the Britannia Bridge. Considering the Almö Bridge, the fact did become completely clear that it was only a matter of time before it was to be hit by a ship adrift!

If we finally do compare the slenderness of the Almö Bridge arches, with the same of the cylinder that we have studied earlier, we find that the difference is not so big. The cross-section of the arches has a diameter of 3.8 meter, and a wall thickness of 22 mm at the base, and 14 mm at the crown (Eqs. 5.28, 5.29):

$$\left(\frac{r}{t}\right)_{max} = \frac{1.9}{0.014} = \underline{135.7} \quad \text{(The Almö Bridge)} \tag{5.28}$$

$$\frac{r}{t} = \frac{32.5}{0.18} = \underline{180.6} \quad \text{(The cylinder)} \tag{5.29}$$

We see from the results above that it is not far from that we can compare the slenderness of the arch profile (at the crown) with that of the cylinder. We can also calculate the equivalent wall thickness of the arch, so that the slenderness becomes the same as for the cylinder (Eq. 5.30):

$$t_{eq} = \frac{1.9}{180.6} = \underline{10.5 \text{ mm}} \tag{5.30}$$

A wall thickness of 10.5 mm – which is not far from the actual value of 14 mm – would thus have given the same slenderness for the arch as for the cylinder! Alternatively we could calculate the equivalent wall thickness of the cylinder, so that the slenderness becomes the same as for the arches of the Almö Bridge (Eq. 5.31):

$$t_{eq} = \frac{180.6}{135.7} \cdot 0.18 = \underline{0.24 \text{ mm}} \tag{5.31}$$

Thus the tube section of the Almö Bridge arches was only 33% thicker in relation to the cylinder, so it is easy to see why the bridge was doomed to fail, not only because of the narrow channel, but also because of the low slenderness. If one had considered this potential hit damage risk better, it would for sure have resulted in thicker walls in combination with longitudinal stiffeners inside the tubes. Alternatively, which also is a better idea; one could have done what prof. Em. Bo Edlund suggested in an article in the Göteborgs-Posten 2005 about the Almö Bridge collapse, and that was to provide the sailing route through the channel with especially designed barriers, which would have prevented the bridge from being hit. Nevertheless, the collapsed bridge was replaced with a modern stay-cable bridge (having a box-girder cross-section), which has given a free sailing height over the entire width of the sound – just as for the Britannia Bridge.

Examples

In example 1–4, we follow one and the same plate, either stiffened or unstiffened, and we see how the capacity with respect to axial loading is affected by the stiffening. In example 5–11, we do the same for a simply supported I-girder, but with respect to bending. The last and final example (no. 12) is a continuous I-girder, and where the shear buckling resistance is checked with some different expressions than for the I-girder studied before.

Example I

The maximum load-carrying capacity of an axially loaded and unstiffened plate is to be calculated. The plate has a width of 500 mm, a length of 2000 mm, and a thickness of 6 mm. Steel quality: $f_y = 275$ MPa.

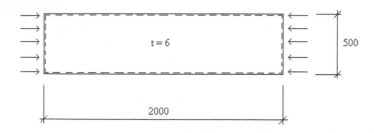

Message

We follow an imaginary *load test*, where the load is raised step by step. First we calculate the critical buckling load, and there-after we find the maximum load-carrying capacity in the ultimate limit state, taking the post-critical reserve effects into account. Worth noticing is the magnitude of the capacities.

Cross-section class

Class 3:
$$\frac{b}{t} \leq 42 \cdot \sqrt{\frac{235}{275}} = \underline{\underline{38.8}}$$

Actual slenderness:
$$\frac{b}{t} = \frac{500}{6} = \underline{\underline{83.3}} \quad \Rightarrow \quad \text{Class 4}$$

Critical buckling load

$$\frac{a}{b} \geq 1 \quad \Rightarrow \quad k = 4$$

$$\sigma_{cr} = 4 \cdot \frac{\pi^2 \cdot 210{,}000}{12 \cdot (1 - 0.3^2) \cdot \left(\dfrac{500}{6}\right)^2} = \underline{\underline{109.4 \, \text{MPa}}}$$

$$\Rightarrow \quad N_{cr} = 109.4 \cdot 500 \cdot 6 = \underline{\underline{328.2 \, \text{kN}}}$$

According to the classical theory, the plate will buckle at a load of rather impressive 33 tons (at the so-called bifurcation point). However, due to the presence of initial imperfections and residual stresses, the true buckling load will occur at a somewhat lower value though.

Maximum load

$$\bar{\lambda}_p = \sqrt{\frac{f_y}{\sigma_{cr}}} = \sqrt{\frac{275}{109.4}} = \underline{\underline{1.585}} \quad (>0.673)$$

$$\rho = \frac{(1.585 - 0.22)}{1.585^2} = \underline{\underline{0.544}} \quad (\psi = 1)$$

$$b_{eff} = 0.544 \cdot 500 = \underline{\underline{272 \, \text{mm}}}$$

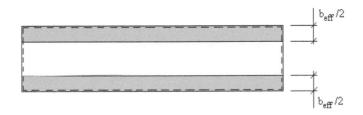

$$\Rightarrow \quad N_{c,Rd} = \frac{A_{eff} \cdot f_y}{\gamma_{M0}} = \frac{272 \cdot 6 \cdot 275}{1.0} = \underline{\underline{448.8 \, \text{kN}}}$$

The capacity has accordingly risen from the buckling load of 33 tons (or lower), to 45 tons, i.e. by having post-critical reserve strength of more than 12 tons.

Example 2

We calculate the capacity once again for the plate given in the first example, however, now stiffened in both longitudinal and transverse direction. We start by only having transverse stiffeners (each 500 millimeters), and then we also add a centrically located stiffener in the longitudinal direction. We check the capacities of the stiffeners in the next-coming example, but in this example we assume them to be strong and rigid enough so that they function as fixed nodal lines at buckling of the plate.

a)

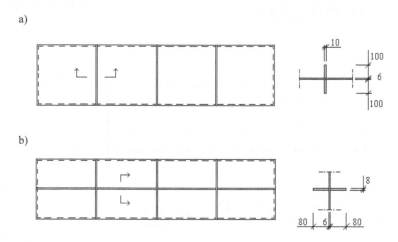

b)

Message

Stiffeners can greatly enhance the load-carrying capacity if they are correctly positioned.

---

a) Without a longitudinal stiffener

Critical buckling load

The stiffeners have divided the plate into four equally large square elements, and as the panel aspect ratio $a/b = 1$ it does not change the critical buckling load – the buckling coefficient, k, is still equal to 4. Just as for the unstiffened plate, the buckles will form within these square panels, with or without stiffeners:

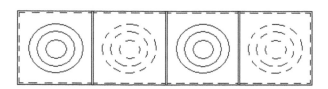

$$\sigma_{cr} = \underline{\underline{109.4\,\text{MPa}}} \quad \Rightarrow \quad N_{cr} = \underline{\underline{328.2\,\text{kN}}} \quad \text{(as before)}$$

If the stiffeners instead had been positioned closer than the width, then the panel aspect ratio a/b had become less than one, which had resulted in an increased buckling coefficient.

Maximum load

Also for the maximum capacity in the ultimate limit state there will be no change. The effective width is still the same, as the critical buckling stress has not changed:

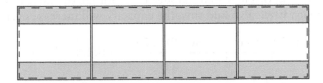

$$b_{eff} = 272\,\text{mm} \quad \Rightarrow \quad N_{c.Rd} = 448.8\,\text{kN}$$

b) With a longitudinal stiffener

Cross-section class

Class 3: $\quad\quad\quad\quad \dfrac{b}{t} \le 38.8 \quad$ (as before)

Actual slenderness: $\dfrac{b}{t} = \dfrac{250}{6} = 41.7 \quad \Rightarrow \quad$ Class 4

Critical buckling load

$$\frac{a}{b} = \frac{500}{250} = 2.0 \quad \Rightarrow \quad k = 4$$

$$\sigma_{cr} = 4 \cdot \frac{\pi^2 \cdot 210{,}000}{12 \cdot (1 - 0.3^2) \cdot \left(\dfrac{250}{6}\right)^2} = 436.6\,\text{MPa} \quad (>275\,\text{MPa})^*$$

*No buckles will here be shown before the maximum capacity is reached, but the plate has still to be designed using an effective width due to the high slenderness (Class 4).

Maximum load

$$\bar{\lambda}_p = \sqrt{\frac{275}{436.6}} = 0.794 \quad (>0.673)$$

$$\rho = \frac{(0.794 - 0.22)}{0.794^2} = 0.910 \quad \Rightarrow \quad b_{eff} = 0.910 \cdot 250 = 227.5\,\text{mm}$$

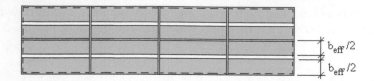

$$\Rightarrow \quad N_{c \cdot Rd} = \frac{2 \cdot 227.5 \cdot 6 \cdot 275}{1.0} = \underline{\underline{750.8 \, \text{kN}}}$$

This capacity is for the original plate alone, i.e. without the additional area of the longitudinal stiffener. If we also add this, the capacity will increase even more:

$$\Rightarrow \quad N'_{c \cdot Rd} = \frac{(2 \cdot 227.5 \cdot 6 + 2 \cdot 80 \cdot 8) \cdot 275}{1.0} = \underline{\underline{1102.8 \, \text{kN}}}$$

The maximum capacity has consequently risen almost 2.5 times in relation to the unstiffened plate!

Example 3

Here we check the capacity of the chosen longitudinal stiffener:

Message

We assume here that the transverse stiffeners are functioning as nodal lines at (global) buckling of the longitudinal stiffener. If it is shown that the capacity of the longitudinal stiffener has to be reduced with respect to buckling – despite the short distance between the transverse stiffeners – then the capacity of the plate will also be reduced.

The plate with the longitudinal stiffener, seen from the loaded edge:

The cross-section to check, with respect to global buckling in between the transverse stiffeners:

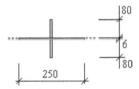

Normal force capacity

$$A_{brutto} = 250 \cdot 6 + 2 \cdot 80 \cdot 8 = 2780 \, \text{mm}^2$$

$$A_{eff} = 227.5 \cdot 6 + 2 \cdot 80 \cdot 8 = 2645 \, \text{mm}^2$$

$$\Rightarrow \quad \beta_A = \frac{A_{eff}}{A_{brutto}} = \frac{2645}{2780} = 0.95$$

$$I = \frac{250 \cdot 6^3}{12} + 2 \cdot \frac{8 \cdot 80^3}{12} + 2 \cdot 8 \cdot 80 \cdot \left(\frac{6}{2} + \frac{80}{2}\right)^2 = 3.054 \cdot 10^6 \, \text{mm}^4$$

$$i = \sqrt{\frac{I}{A}} = \sqrt{\frac{3.054 \cdot 10^6}{2780}} = 33.1 \, \text{mm}$$

$$\lambda = \frac{l_c}{i} = \frac{500}{33.1} = \underline{15.1} \quad \lambda_1 = 93.9 \cdot \sqrt{\frac{235}{275}} = \underline{86.4}$$

$$\Rightarrow \quad \bar{\lambda} = \frac{\lambda}{\lambda_1} \cdot \sqrt{\beta_A} = \frac{15.1}{86.4} \cdot \sqrt{0.95} = \underline{0.17}$$

As the reference value of the slenderness does not exceed 0.2, it means that the capacity of the plate becomes unreduced with respect to the buckling strength of the longitudinal stiffener.

Refined model

As an alternative to the simplified buckling model used above, there is a refined model to be used, where the web plate is regarded as an "elastic bed", that is supporting the stiffener against out-of-plane buckling (see section 4.2 pages 75–77). We check the capacity for the effective net section:

$$A_{eff} = 227.5 \cdot 6 + 2 \cdot 8 \cdot 80 = \underline{2642\,\text{mm}^2}$$

$$I_{st} = \frac{227.5 \cdot 6^3}{12} + 2 \cdot \frac{8 \cdot 80^3}{12} + 2 \cdot 8 \cdot 80 \cdot \left(\frac{6}{2} + \frac{80}{2}\right)^2 = \underline{3.053 \cdot 10^6\,\text{mm}^4}$$

$$a_c = 4.33 \cdot \sqrt[4]{\frac{3.053 \cdot 10^6 \cdot 250^2 \cdot 250^2}{6^3 \cdot 500}} = \underline{576\,\text{mm}} \quad (>500\,\text{mm})$$

$$N_{cr} = \frac{\pi^2 \cdot 210{,}000 \cdot 3.053 \cdot 10^6}{500^2} + \frac{210{,}000 \cdot 6^3 \cdot 500 \cdot 500^2}{35.7 \cdot 250^2 \cdot 250^2}$$

$$= 25\,310.8 + 40.7 = \underline{25\,351.5\,\text{kN}}$$

$$\lambda_c = \sqrt{\frac{A_{eff} \cdot f_y}{N_{cr}}} = \sqrt{\frac{2645 \cdot 275}{25\,351.5 \cdot 10^3}} = \underline{0.17} \quad (<0.2)$$

The result here also showed that no reduction with respect to buckling of the stiffener has to be made (which also was a rather expected result, as it was a matter of a refined model in relation the first more simplified approach). It can be noted, from the results above, that the stiffening effect from the web plate, acting as a transverse bed, was more or less negligible. The plate bending stiffness of the web plate was apparently rather small in comparison to the bending stiffness of the stiffener.

Example 4

Here we finally check the capacity of the transverse stiffeners:

Message

We check what we have assumed in the former example, and that was that the transverse stiffeners were functioning as fixed nodal lines for the buckling of the longitudinal stiffener.

The transverse stiffeners are designed according section 4.2 page 77, with respect to a required stiffness in order to balance the inclined force coming from the deviating out-of-straight longitudinal stiffener – a normal value of this deviation is set to 2%:

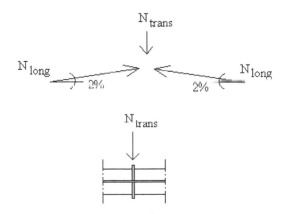

The transverse stiffener shall deflect less than $b/500$ in the load application point because of this force, and the cross-section to use is, beside the transverse stiffeners on either of the web, also an interacting part of the web, normally set to $40 \cdot t_w$:

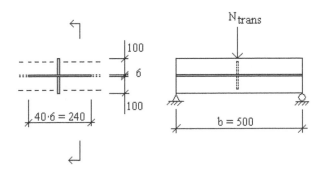

$$I = \frac{240 \cdot 6^3}{12} + 2 \cdot \frac{10 \cdot 100^3}{12} + 2 \cdot 10 \cdot 100 \cdot \left(\frac{6}{2} + \frac{100}{2}\right)^2$$

$$= \underline{\underline{7.20 \cdot 10^6 \, \text{mm}^4}}$$

$$N_{long} = \frac{A_{eff} \cdot f_y}{\gamma_{M1}} = \frac{2645 \cdot 275}{1.0} = \underline{\underline{727.4 \, \text{kN}}}$$

$$N_{trans} = 0.02 \cdot 727.4 = \underline{\underline{14.5 \, \text{kN}}}$$

$$\delta = \frac{N_{trans} \cdot b^3}{48 \cdot EI_{trans}} = \frac{14.5 \cdot 10^3 \cdot 500^3}{48 \cdot 210,000 \cdot 7.29 \cdot 10^6} = \underline{\underline{0.02 \, \text{mm}}}$$

$$\delta \le \frac{b}{500} = \frac{500}{500} = \underline{\underline{1 \, \text{mm}}} \quad \text{OK!}$$

The chosen stiffener in the transverse direction met the requirement regarding its stiffness without any problem, and by doing so, it also means that the load-carrying capacity (with respect to stress) is sufficient.

Example 5

The maximum design bending moment resistance, $M_{b,Rd}$, is to be determined for this double-symmetrical and unstiffened I-girder. This is done by in turn determining cross-section class, effective width, position of the neutral axis for the net-section, and finally the design moment resistance. $f_y = 355$ MPa

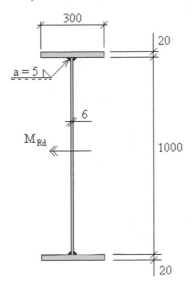

Message

A substantial part of the load-carrying capacity still remains after that local buckling has occurred (at $\sigma_{cr} = 168$ MPa), and this is because of the post-critical reserve strength. The maximum load-carrying capacity is consequently reached first after buckling, when yielding is reached in the outermost fibre (i.e. $\sigma_{max} = f_y$).

Cross-section class

Cross-section class for the flange:

Class 1: $\quad \dfrac{c}{t_f} \leq 9 \cdot \varepsilon = 9 \cdot \sqrt{\dfrac{235}{f_y}} = 9 \cdot \sqrt{\dfrac{235}{355}} = 9 \cdot 0.81 = \underline{\underline{7.3}}$

Actual slenderness: $\quad \dfrac{c}{t_f} = \dfrac{\dfrac{300}{2} - \dfrac{6}{2} - \sqrt{2} \cdot 5}{20} = \underline{\underline{7.0}}$ OK!

Cross-section class for the web:

Class 3: $\quad \dfrac{d}{t_w} \leq 124 \cdot \varepsilon = 124 \cdot 0.81 = \underline{\underline{100.4}}$

Actual slenderness: $\dfrac{d}{t_w} = \dfrac{1000 - 2 \cdot \sqrt{2} \cdot 5}{6} = \underline{\underline{164.3}}$ $\Rightarrow$ Class 4

Effective cross-section

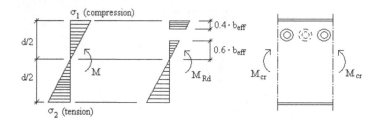

$d = 1000 - 2 \cdot \sqrt{2} \cdot 5 = \underline{\underline{986\,\text{mm}}}$

$b_{eff} = \rho \cdot \dfrac{d}{2}$

$\psi = \dfrac{\sigma_2}{\sigma_1} = -1$ $\Rightarrow$ $k_\sigma = \underline{\underline{23.9}}$

$\sigma_{cr} = 23.9 \cdot \dfrac{\pi^2 \cdot 210{,}000}{12 \cdot (1 - 0.3^2) \cdot \left(\dfrac{986}{6}\right)^2} = \underline{\underline{168.0\,\text{MPa}}}$

$\bar{\lambda}_p = \sqrt{\dfrac{f_y}{\sigma_{cr}}} = \sqrt{\dfrac{355}{168}} = \underline{\underline{1.454}}$ (>0.673)

$\rho = \dfrac{(1.454 - 0.22)}{1.454^2} = \underline{\underline{0.58}}$

$b_{eff} = \rho \cdot b_c = \rho \cdot \dfrac{d}{2} = 0.58 \cdot \dfrac{986}{2} = \underline{\underline{286\,\text{mm}}}$

$^1\,0.4 \cdot 286 + \sqrt{2} \cdot 5 = \underline{\underline{121\,\text{mm}}}$

$^2\,\dfrac{986}{2} - 286 = \underline{\underline{207\,\text{mm}}}$

$^3\,0.6 \cdot 286 = \underline{\underline{172\,\text{mm}}}$

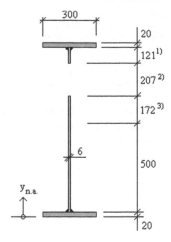

Area of the effective net-section:

$$A_{eff} = 2 \cdot 20 \cdot 300 + (500 + 172 + 121) \cdot 6 = \underline{\underline{16758 \, \text{mm}^2}}$$

Neutral axis of the effective net-section:

$$y_{n.a.} \cdot A_{eff} = 20 \cdot 300 \cdot \frac{20}{2} + (500 + 172) \cdot 6 \cdot \left(20 + \frac{672}{2}\right)$$

$$+ 121 \cdot 6 \cdot \left(1020 - \frac{121}{2}\right) + 20 \cdot 300 \cdot \left(1020 + \frac{20}{2}\right)$$

$$= \underline{\underline{8371\,989 \, \text{mm}^3}}$$

$$\Rightarrow \quad y_{n.a.} = \frac{8371\,989}{16758} = \underline{\underline{500 \, \text{mm}}}$$

The neutral axis has thus been lowered 20 mm (520–500) for the effective net-section in relation to the gross-section.

Bending moment resistance

$$I_{eff} = 20 \cdot 300 \cdot \left(500 - \frac{20}{2}\right)^2 + \frac{6 \cdot 672^3}{12} + 6 \cdot 672 \cdot \left(480 - \frac{672}{2}\right)^2$$

$$+ \frac{6 \cdot 121^3}{12} + 6 \cdot 121 \cdot \left(520 - \frac{121}{3}\right)^2 + 20 \cdot 300 \cdot \left(1020 + \frac{20}{2} - 500\right)^2$$

$$= \underline{\underline{35.16 \cdot 10^8 \, \text{mm}^4}}$$

$$W_{eff}^c = \frac{I_{eff}}{\Delta y_c} = \frac{35.16 \cdot 10^8}{(1040 - 500)} = \underline{\underline{65.11 \cdot 10^5 \text{ mm}^3}}$$

$$W_{eff}^t = \frac{I_{eff}}{\Delta y_t} = \frac{35.16 \cdot 10^8}{500} = \underline{\underline{70.32 \cdot 10^5 \text{ mm}^3}}$$

Due to the position of the neutral axis, yielding will first occur at the top compression side, and therefore the bending moment resistance is determined with respect to the corresponding section modulus (i.e. the lowest value):

$$M_{c,Rd} = \frac{W_{eff}^c \cdot f_y}{\gamma_{M0}} = \frac{65.11 \cdot 10^{-4} \cdot 355 \cdot 10^3}{1.0} = \underline{\underline{2311 \text{ kNm}}}$$

Example 6

In this example we seek the bending moment resistance of the cross-section in example 5, however, now strengthened with a longitudinal stiffener in the upper compression part of the web, positioned 200 millimeters below the lower face of the upper flange (approximately where the maximum amplitude of the normal stress buckles could be expected). The stiffener is fillet welded to the web plate, with a throat thickness $a = 5$ mm. We assume that the longitudinal stiffener is sufficiently strong so that global buckling of the same (in between the transverse stiffeners) does not reduce the bending moment resistance of the cross-section of the girder (we check this assumption in example 7).

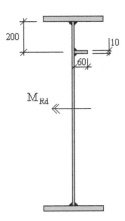

Message

By adding stiffeners where we expect the maximum amplitude of the buckles to appear (for the unstiffened case) – and by doing so, creating a nodal line there – we increase the bending moment resistance of the girder.

The position of the stiffener:

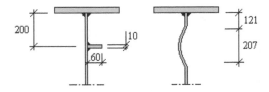

Longitudinal stiffeners are normally, as a rule, positioned $0.2 \cdot d$ from the edge of the web plate, but in our case, the more exact position would have been:

$$121 + \frac{207}{2} = \underline{\underline{225 \, \text{mm}}}$$

In the continued analysis, we proceed with the given position, i.e. 200 mm $(0.2 \cdot 1000)$.

Cross-section class

Longitudinal stiffener

Class 1: $\dfrac{c}{t_f} \le 9 \cdot \varepsilon = 9 \cdot 0.81 = \underline{\underline{7.3}}$

Actual slenderness: $\dfrac{c}{t} = \dfrac{60 - \sqrt{2} \cdot 5}{10} = \underline{\underline{5.3}}$ OK!

The longitudinal stiffener is so compact that it can fully plastify prior to any local buckling.

The upper part of the web plate

Class 1: $\dfrac{d}{t_w} \le 33 \cdot \varepsilon = 33 \cdot 0.81 = \underline{\underline{26.7}}$

Class 2: $\dfrac{d}{t_w} \le 38 \cdot \varepsilon = 38 \cdot 0.81 = \underline{\underline{30.8}}$

Actual slenderness: $\dfrac{d}{t_w} = \dfrac{200 - \dfrac{10}{2} - 2 \cdot 5 \cdot \sqrt{2}}{6} = \underline{\underline{30.1}}$

As the actual slenderness of the upper part of the web plate fulfills the requirement of Class 2, it has the capacity to fully plastify:

The lower part of the web plate

In the following we neglect the small effect that the stiffener has on the position of the neutral axis (on the safe side as the neutral axis will be somewhat lowered, and the compression zone will decrease):

Class 1: $\dfrac{d}{t_w} \le \dfrac{36 \cdot \varepsilon}{\alpha} = \dfrac{36 \cdot 0.81}{0.375} = \underline{\underline{78}}$

Class 2: $\dfrac{d}{t_w} \le \dfrac{41.5 \cdot \varepsilon}{\alpha} = \dfrac{41.5 \cdot 0.81}{0.375} = \underline{\underline{90}}$

Class 3: $\dfrac{d}{t_w} \le 62 \cdot \varepsilon \cdot (1 - \psi) \cdot \sqrt{(-\psi)} = 62 \cdot 0.81 \cdot (1 + 1.67) \cdot \sqrt{1.67} = \underline{\underline{173}}$

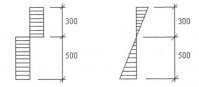

$$\alpha = \frac{300}{800} = 0.375 \quad \psi = \frac{-500}{300} = -1.67$$

(In EC3 compression is given the positive sign)

Actual slenderness: $\dfrac{d}{t_w} = \dfrac{800 - \dfrac{10}{2} - 2 \cdot 5 \cdot \sqrt{2}}{6} = 130.1$

As the lower part of the web plate belongs to Class 3, it tells us that only linear stress distribution is possible (however, with yielding of the outermost compression fibre of this part).

The resulting stress distribution for the entire cross-section:

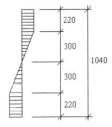

Bending moment resistance

$$W_{el,pl} = 2 \cdot 20 \cdot 300 \cdot \left(500 + \frac{20}{2}\right) + 2 \cdot 6 \cdot 200 \cdot \left(300 + \frac{200}{2}\right)$$

$$+ 2 \cdot \frac{6 \cdot 300}{2} \cdot \frac{2}{3} \cdot 300 = 74.40 \cdot 10^5 \, \text{mm}^3$$

$$\Rightarrow \qquad M_{c,Rd} = \frac{W_{el,pl} \cdot f_y}{\gamma_{M0}} = \frac{74.4 \cdot 10^{-4} \cdot 355 \cdot 10^3}{1.0} = 2641 \, \text{kNm}$$

In comparison to the unstiffened cross-section (in example 5) we have received a 14% increase of the moment resistance. The capacity will increase even more if we also add the capacity coming from the stiffener:

$$\Delta M \approx \frac{A_{st} \cdot \Delta y \cdot f_y}{\gamma_{M0}} = \frac{10 \cdot 60 \cdot 300 \cdot 10^{-9} \cdot 355 \cdot 10^3}{1.0} = 63.9 \, \text{kNm}$$

Example 7

In this example we check the longitudinal stiffener (that was chosen in example 6) with respect to global buckling in between the transverse stiffeners. We assume transverse stiffeners having the dimension of $10 \times 100 \text{ mm}^2$, every meter along the girder, that is simply supported and 6.0 meter long:

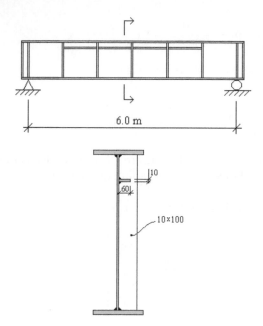

6.0 m

10
60
10×100

Message

Would it show that the chosen stiffeners has a reduced capacity to carry a normal force, then the bending moment resistance of the girder is also affected, and has to be reduced.

Longitudinal stiffener

The interacting effective web area is chosen according to the figure below (see also section 4.2 pages 75–77):

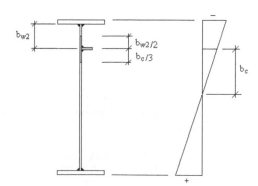

b_{w2}

$b_{w2}/2$
$b_c/3$
b_c

+

with $b_{w2} = \underline{\underline{200}}$ and $b_c = \underline{\underline{300}}$ the cross-section becomes:

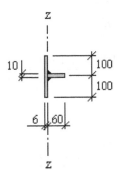

$$A_{long} = 6 \cdot 200 + 10 \cdot 60 = \underline{\underline{1800 \,\text{mm}^2}}$$

$$I_{st}^{z-z} = 6 \cdot 200 \cdot 3^2 + \frac{10 \cdot 60^3}{12} + 10 \cdot 60 \cdot \left(\frac{60}{2}\right)^2$$

$$= \underline{\underline{73.08 \cdot 10^4 \,\text{mm}^4}}$$

$$a_c - 4.33 \cdot \sqrt[4]{\frac{I_{st} \cdot b_{w1}^2 \cdot b_{w2}^2}{t_w^3 \cdot b_w}}$$

$$a_c = 4.33 \cdot \sqrt[4]{\frac{73.08 \cdot 10^4 \cdot 800^2 \cdot 200^2}{6^3 \cdot 1000}} = \underline{\underline{2349 \,\text{mm}}}$$

As the buckling length of the longitudinal stiffener does exceed the distance between the transverse stiffeners ($a = 1000$ mm), the critical buckling load becomes:

$$N_{cr} = \frac{\pi^2 \cdot EI_{st}}{a^2} + \frac{E \cdot t_w^3 \cdot b_w \cdot a^2}{35.7 \cdot b_{w1}^2 \cdot b_{w2}^2}$$

$$= \frac{\pi^2 \cdot 2.1 \cdot 10^5 \cdot 73.08 \cdot 10^4}{1000^2} + \frac{2.1 \cdot 10^5 \cdot 6^3 \cdot 1000 \cdot 1000^3}{35.7 \cdot 800^2 \cdot 200^2}$$

$$= 1515 + 50 = \underline{\underline{1565 \, \text{kN}}}$$

$$\lambda_c = \sqrt{\frac{A_{eff} \cdot f_{yk}}{N_{cr}}} = \sqrt{\frac{1800 \cdot 355}{1565 \cdot 10^3}} = \underline{\underline{0.639}}$$

$$\lambda_c = \underline{\underline{0.639}} \quad \text{and curve } c \quad \Rightarrow \quad \chi = \underline{\underline{0.754}}$$

The load-carrying capacity of the longitudinal stiffener – with respect to the buckling resistance – has to be reduced, and consequently also the bending moment resistance of the girder:

$$M'_{c,Rd} \approx 0,754 \cdot 2641 = \underline{\underline{1991 \, \text{kNm}}}$$

As the reduced value of the moment resistance is below what has been calculated for the unstiffened girder (example 5; $M_{c,Rd} = 2311$ kNm), it shows that the assumed design model is too much on the safe side. Even if buckling of the stiffener is a global phenomenon, there are post-critical reserve effects that have been neglected. The true moment resistance of the stiffened girder must be in between the upper and lower limits of what has been calculated earlier, i.e. $2311 \leq M'_{c,Rd} \leq 2641$ kNm.

Transverse stiffeners

The condition for the calculated buckling resistance of the longitudinal stiffener is that the transverse stiffeners must have a minimum stiffness so that they function as fixed nodal lines at buckling of the longitudinal stiffener. They shall deflect less than $b_w/500$ for a force equal to 2% of the axial force in the longitudinal stiffener (compare section 4.2 page 77, and example 4):

$$A_{trans} = 240 \cdot 6 + 10 \cdot 100 = \underline{\underline{2440 \, \text{mm}^2}}$$

$$N_{long} = N_{b,Rd} = \frac{\chi \cdot \beta_A \cdot A_{long} \cdot f_y}{\gamma_{M1}}$$

$$= \frac{0.754 \cdot 1 \cdot 1800 \cdot 355}{1.0} = \underline{\underline{481.8 \, \text{kN}}}$$

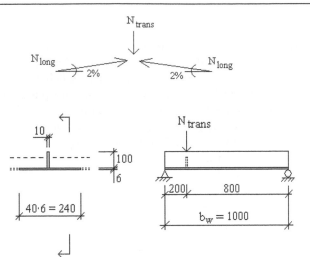

$$N_{trans} = 0.02 \cdot 481.8 = \underline{\underline{9.6\,\text{kN}}}$$

$$y_{n.a.} = \frac{240 \cdot 6 \cdot \dfrac{6}{2} + 10 \cdot 100 \cdot \left(6 + \dfrac{100}{2}\right)}{2440} = \underline{\underline{25\,\text{mm}}}$$

$$I_{trans} = \frac{240 \cdot 6^3}{12} + 240 \cdot 6 \cdot \left(25 - \frac{6}{2}\right)^2 + \frac{10 \cdot 100^3}{12}$$

$$+ 10 \cdot 100 \cdot \left(6 + \frac{100}{2} - 25\right)^2 = \underline{\underline{24.96 \cdot 10^5\,\text{mm}^4}}$$

$$\delta = \frac{N_{trans} \cdot b_{w1}^2 \cdot b_{w2}^2}{3 \cdot EI_{trans} \cdot b_w}$$

$$= \frac{9.6 \cdot 10^3 \cdot 800^2 \cdot 200^2}{3 \cdot 2.1 \cdot 10^5 \cdot 24.96 \cdot 10^5 \cdot 1000} = \underline{\underline{0.15\,\text{mm}}}$$

$$\delta_{till} = \frac{b_w}{500} = \frac{1000}{500} = \underline{\underline{2\,\text{mm}}} \quad \text{OK!}$$

The capacity of the longitudinal stiffener and the cross-section resistance of the girder need not be further reduced with respect to insufficient stiffness of the transverse stiffeners.

Example 8

In this example we check the girder with respect to the lateral/torsional buckling risk. Even if this is a global instability phenomenon, which not can be related to the local buckling risk of the cross-section of the girder (which also is the focus of this book), we carry out such a check, as this is a very important phenomenon that *never* should be neglected. We assume the girder to be stabilized laterally at the supports (by double-sided transverse stiffeners) and at the load application point in mid-span (lateral support of the top compression flange). We assume the self-weight of the girder to be included in the applied load.

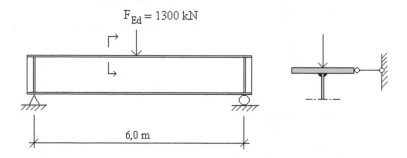

Message

Lateral/torsional buckling is normally quite complicated to check (and can thus "hide" the phenomenon for the designer), however, in this example we perform the analysis in a simplified manner (on the safe side), where we regard the top compression flange as an isolated element, that is buckling in the lateral transverse direction in between the supports and the load application point.

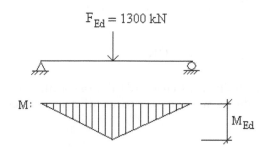

$$M_{Ed} = \frac{F_{Ed} \cdot L}{4} = \frac{1300 \cdot 6.0}{4} = \underline{\underline{1950\,kNm}}$$

We check the upper compression flange for *global* buckling in the *transverse direction*:

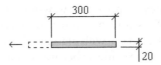

$$A_f = 20 \cdot 300 = \underline{\underline{6000\,\text{mm}^2}}$$

$$I_f = \frac{20 \cdot 300^3}{12} = \underline{\underline{45 \cdot 10^6\,\text{mm}^4}}$$

We assume the buckling length being equal to 3.0 meter, despite the fact that the axial force – which is following the moment distribution – is not constant (and therefore should have given a somewhat lower buckling length). A view of the buckled upper flange seen from above:

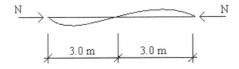

$$N_{cr} = \frac{\pi^2\,EI_f}{L_c^2} = \frac{\pi^2 \cdot 2.1 \cdot 10^5 \cdot 45 \cdot 10^6}{(3.0 \cdot 10^3)^2} = \underline{\underline{10,363\,kN}}$$

$$\bar{\lambda} = \sqrt{\frac{\beta_A \cdot A \cdot f_y}{N_{cr}}} = \sqrt{\frac{1.0 \cdot 6000 \cdot 355}{10,363 \cdot 10^3}} = \underline{\underline{0.45}}$$

$$\bar{\lambda} = 0.45 \text{ and buckling curve } c \quad \Rightarrow \quad \chi = \underline{\underline{0.87}}$$

$$M_{c,Rd} = 0.87 \cdot 2311 = \underline{\underline{2011\,\text{kNm}}} \text{ (unstiffened girder; example 5)}$$

$$M_{c,Rd} = 0.87 \cdot 2641 = \underline{\underline{2298\,\text{kNm}}} \text{ (stiffened girder; example 6)}$$

Both the stiffened and the unstiffened girder meet the load-carrying requirement with respect to the lateral/torsional buckling risk.

Example 9

If we take the cross-section that we have studied in example 5–8 (so far), and transform the web plate so that it becomes compact, how would this change affect the stiffness and load-carrying capacity? We study two alternative cross-sections, where we have kept the original flanges, but altered the web plate. In the first cross-section (A), the web area – and also the total area – is the same as before, and in the second cross-section (B), the web area has been increased with 60%.

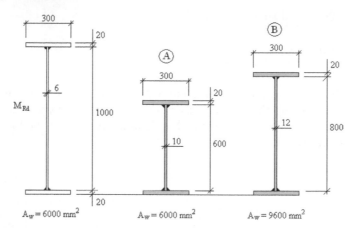

Message

The only reason to choose a compact cross-section (besides a need for robustness and stability during transport and handling) is if the statical system, and the safety with respect to local buckling, are such that there is a need for plastic redistribution of moments, so that a mechanism can be formed in the ultimate limit state by plastic hinges having a sufficient rotational capacity. Normally though, there is a loss of both stiffness and load-carrying capacity of choosing such a compact cross-section in relation to a thin-walled alternative.

In the following analysis we assume that both alternatives are in class 2 or better, with respect to the slenderness of the web plate, so that there is a full capacity to plastify without any risk of buckling.

Cross-section A

$$M_{pl \cdot Rd} = \frac{W_{pl} \cdot f_y}{\gamma_{M0}}$$

$$W_{pl} = 2 \cdot 20 \cdot 300 \cdot 310 + 2 \cdot 10 \cdot 300 \cdot 150 = \underline{\underline{4.62 \cdot 10^6 \, \text{mm}^3}}$$

$$\Rightarrow \quad M_{pl,Rd} = \frac{4.62 \cdot 10^{-3} \cdot 355 \cdot 10^3}{1.0} = \underline{\underline{1640 \, \text{kNm}}}$$

$$M_{c,Rd} = \underline{\underline{2311 \, \text{kNm}}} \quad \text{(compare example 5)}$$

Also the stiffness, and consequently the deflections, are affected negatively:

$$I_A = 2 \cdot 20 \cdot 300 \cdot 310^2 + \frac{10 \cdot 600^3}{12} = \underline{\underline{13.3 \cdot 10^8 \, \text{mm}^4}}$$

$$I = \underline{\underline{36.2 \cdot 10^8 \, \text{mm}^4}} \quad \text{(gross-section example 5)}$$

Cross-section B

$$W_{pl} = 2 \cdot 20 \cdot 300 \cdot 410 + 2 \cdot 12 \cdot 400 \cdot 200 = \underline{\underline{6.84 \cdot 10^6 \, \text{mm}^3}}$$

$$\Rightarrow \quad M_{pl.Rd} = \frac{6.84 \cdot 10^{-3} \cdot 355 \cdot 10^3}{1.0} = \underline{\underline{2428 \, \text{kNm}}}$$

$$M_{c,Rd} = \underline{\underline{2311 \, \text{kNm}}} \quad \text{(compare example 5)}$$

Even if the moment resistance of alternative B slightly exceeds the original capacity, the stiffness is still lower (compared to the value given above):

$$I_B = 2 \cdot 20 \cdot 300 \cdot 410^2 + \frac{12 \cdot 800^3}{12} = \underline{\underline{25.3 \cdot 10^8 \, \text{mm}^4}}$$

Alternative B gave a load-carrying capacity that was comparable to the original cross-section, but the stiffness was still lower. But not to forget; cross-section B is definitely a more costly alternative, as the web area is 60% larger than the original (20% if the entire areas are compared).

Example 10

In this example we check the girder for the concentrated transverse force $F_{Ed} = 1300$ kN, that is positioned at the top flange in mid-span. The force is distributed over an area of $200 \times 300 \text{ mm}^2$, through a bearing plate. We check two alternatives; with or without vertical web stiffeners under the load.

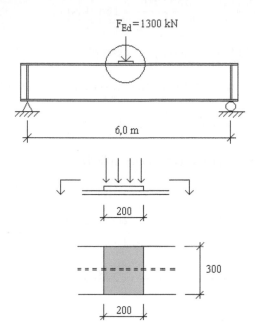

Message

As a general rule, vertical web stiffeners are always positioned where local concentrated loads are introduced (despite the actual need according to an analysis), however, in this example we still do start with a check of the unstiffened case.

$$F_{Rd} = \frac{f_{yw} \cdot L_{eff} \cdot t_w}{\gamma_{M1}}$$

$$L_{eff} = \chi_F \cdot l_y$$

$$\chi_F = \frac{0.5}{\overline{\lambda}_F} \quad (\leq 1.0)$$

$$\overline{\lambda}_F = \sqrt{\frac{l_y \cdot t_w \cdot f_{yw}}{F_{cr}}}$$

$$F_{cr} = 0.9 \cdot k_F \cdot E \cdot \frac{t_w^3}{h_w}$$

$$k_F = 6 + 2 \cdot \left(\frac{h_w}{a}\right)^2 = 6 + 2 \cdot \left(\frac{1.0}{6.0}\right)^2 = \underline{\underline{6.056}}$$

$$F_{cr} = 0.9 \cdot 6.056 \cdot 2.1 \cdot 10^8 \cdot \frac{(6 \cdot 10^{-3})^3}{1.0} = \underline{\underline{247.2\,\text{kN}}}$$

$$l_y = s_s + 2 \cdot t_f \cdot (1 + \sqrt{m_1 + m_2}) \quad (\leq a)$$

$$m_1 = \frac{f_{yf} \cdot b_f}{f_{yw} \cdot t_w} = \frac{355 \cdot 300}{355 \cdot 6} = \underline{\underline{50}}$$

$$m_2 = 0.02 \cdot \left(\frac{h_w}{t_f}\right)^2 = 0.02 \cdot \left(\frac{1000}{20}\right)^2 = \underline{\underline{50}} \quad (\bar{\lambda}_F > 0.5)$$

$$s_s = \underline{\underline{200\,\text{mm}}}$$

$$l_y = 200 + 2 \cdot 20 \cdot (1 + \sqrt{50 + 50}) = \underline{\underline{640\,\text{mm}}} \quad (<a)$$

$$\bar{\lambda}_F = \sqrt{\frac{640 \cdot 6 \cdot 10^{-6} \cdot 355 \cdot 10^3}{247.2}} = \underline{\underline{2.348}} \quad (>0.5,\ m_2 = 50)$$

$$\chi_F = \frac{0.5}{2.348} = \underline{\underline{0.213}} \quad (<1.0)$$

$$L_{eff} = 0.213 \cdot 640 = \underline{\underline{136\,\text{mm}}}$$

$$\Rightarrow \quad F_{Rd} = \frac{355 \cdot 10^3 \cdot 136 \cdot 6 \cdot 10^{-6}}{1.0} = \underline{\underline{289.7\,\text{kN}}}$$

$\ll F_{Ed} = 1300\,\text{kN}$ Not OK! The web needs to be stiffened!

The interaction criterion – the combined effect of transverse force and bending moment – has not to be checked, as it is superfluous.

Design of vertical web stiffeners

We check if the assumption of double-sided vertical web stiffeners, having the dimension of $120 \times 15\,\text{mm}^2$, fulfill the requirement regarding a minimum load-carrying capacity of $N_{b,Rd} = 1300\,\text{kN}$. We assume the stiffeners to be in class 3 (or better), which assures us that full plastification without any risk of local buckling is possible.

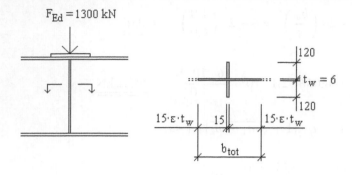

$$b_{tot} = 2 \cdot (15 \cdot 0.81 \cdot 6) + 15 = \underline{\underline{161 \text{ mm}}}$$

$$\bar{\lambda} = \sqrt{\frac{\beta_A \cdot A \cdot f_y}{N_{cr}}}$$

$$A = 2 \cdot 15 \cdot 120 + 161 \cdot 6 = \underline{\underline{4566 \text{ mm}^2}}$$

$$I = \frac{161 \cdot 6^3}{12} + 2 \cdot \frac{15 \cdot 120^3}{12} + 2 \cdot 15 \cdot 120 \cdot \left(3 + \frac{120}{2}\right)^2$$

$$= \underline{\underline{1.861 \cdot 10^7 \text{ mm}^4}}$$

$$L_{cr} = 0.75 \cdot d = 0.75 \cdot 1.0 = \underline{\underline{0.75 \text{ m}}}$$

$$N_{cr} = \frac{\pi^2 \cdot 2.1 \cdot 10^8 \cdot 1.861 \cdot 10^{-5}}{0.75^2} = \underline{\underline{68.6 \text{ kN}}}$$

$$\bar{\lambda} = \sqrt{\frac{1.0 \cdot 4566 \cdot 10^{-6} \cdot 355 \cdot 10^3}{68.6}} = \underline{\underline{0.154}}$$

$$\bar{\lambda} = 0.154 \text{ and buckling curve } c \quad \Rightarrow \quad \chi = \underline{\underline{1.0}}$$

$$N_{b,Rd} = \frac{1.0 \cdot 1.0 \cdot 4566 \cdot 355}{1.0} = \underline{\underline{1621 \text{ kN}}} \quad (>1300 \text{ kN} \quad \text{OK!})$$

The capacity is sufficient first after that stiffeners are put in place!

Example 11

As a last check of the load-carrying capacity of the girder that was introduced in example 5, we calculate the shear buckling resistance, $V_{b,Rd}$. We assume the bending moment resistance of the girder to be unaffected by any lateral/torsional buckling risk (e.g. by having the top compression flange supported along the entire length). The self-weight of the girder can be assumed to be included in the applied load. We also assume that the vertical stiffeners (outside the supports and in between) meet the requirements with respect to resistance and stiffness.

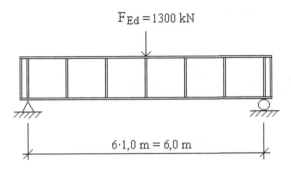

Message

Vertical web stiffeners increase the shear resistance of the web – the diagonal tension field is supported in the vertical direction for its anchorage, and the efficiency of this action is enhanced by a decreased distance between the stiffeners. The design shear panels for the girder above are the two panels in the centre (on either side of the load), and this is due to the maximum shear force in combination with maximum bending moment in these two panels.

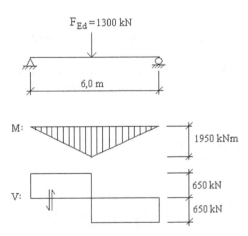

$$\frac{a}{h_w} = \frac{1.0}{1.0} = \underline{\underline{1.0}} \quad \Rightarrow \quad k_\tau = 5.34 + \frac{4.0}{1.0^2} = \underline{\underline{9.34}}$$

$$\tau_{cr} = k_\tau \cdot \frac{\pi^2 \cdot E}{12 \cdot (1 - v^2) \cdot \left(\dfrac{h_w}{t}\right)^2}$$

$$= 9.34 \cdot \frac{\pi^2 \cdot 2.1 \cdot 10^5}{12 \cdot (1 - 0.3^2) \cdot \left(\dfrac{1000}{6}\right)^2} = \underline{\underline{63.8 \, \text{MPa}}}$$

$$\bar{\lambda}_w = 0.76 \cdot \sqrt{\frac{f_{yw}}{\tau_{cr}}} = 0.76 \cdot \sqrt{\frac{355}{63.8}} = \underline{\underline{1.793}}$$

$\bar{\lambda}_w > 1.08$ and inner panels ("rigid end post"):

$$\chi_w = \frac{1.37}{(0.7 + \bar{\lambda}_w)} = \frac{1.37}{(0.7 + 1.793)} = \underline{\underline{0.550}}$$

$$V_{bw,Rd} = \frac{\chi_w \cdot f_{yw} \cdot h_w \cdot t}{\sqrt{3} \cdot \gamma_{M1}}$$

$$= \frac{0.550 \cdot 355 \cdot 10^3 \cdot 1000 \cdot 6 \cdot 10^{-6}}{\sqrt{3} \cdot 1.0} = \underline{\underline{676.4 \, \text{kN}}}$$

$$M_{f,Rd} = \frac{300 \cdot 20 \cdot 1020 \cdot 10^{-9} \cdot 355 \cdot 10^3}{1.0} = \underline{\underline{2172.6 \, \text{kNm}}}$$

(The entire width of the flange is effective, as the outstand width of the flange does not exceed $15 \cdot \epsilon \cdot t_f$)

As the flange only design bending moment of resistance do exceed the design bending moment, $M_{Ed} = 1950 \, \text{kNm}$, there is a contribution to the design shear resistance coming from the flanges:

$$c = a \cdot \left[0.25 + \frac{1.6 \cdot b_f \cdot t_f^2 \cdot f_{yf}}{t \cdot h_w^2 \cdot f_{yw}}\right]$$

$$= 1000 \cdot \left[0.25 + \frac{1.6 \cdot 300 \cdot 20^2 \cdot 355}{6 \cdot 1000^2 \cdot 355}\right] = \underline{\underline{282 \, \text{mm}}}$$

$$V_{bf,Rd} = \frac{b_f \cdot t_f^2 \cdot f_{yf}}{c \cdot \gamma_{M1}} \cdot \left[1 - \left(\frac{M_{Ed}}{M_{f,Rd}} \right)^2 \right]$$

$$= \frac{300 \cdot 20^2 \cdot 10^{-9} \cdot 355 \cdot 10^3}{282 \cdot 10^{-3} \cdot 1.0} \cdot \left[1 - \left(\frac{1950}{2172.6} \right)^2 \right] = \underline{\underline{29.4 \, \text{kN}}}$$

The shear buckling design resistance finally becomes:

$$V_{b,Rd} = V_{bw,Rd} + V_{bf,Rd} = 676.4 + 29.4 = \underline{\underline{705.8 \, \text{kN}}} \quad (> V_{Ed} \quad \text{OK!})$$

$$\left(< \frac{\eta \cdot f_{yw} \cdot h_w \cdot t}{\sqrt{3} \cdot \gamma_{M1}} = \frac{1.20 \cdot 355 \cdot 10^3 \cdot 1000 \cdot 6 \cdot 10^{-6}}{\sqrt{3} \cdot 1.0} = \underline{\underline{1475.7 \, \text{kN}}} \right)$$

Example 12

In this example we check the shear buckling resistance, $V_{b,Rd}$, of a continuous girder in two spans – the same web dimension but *reduced flanges* in comparison to the girder used before. The vertical stiffeners are positioned with an equal spacing of $a = 1.5$ m, and they can be assumed to be strong enough to carry the vertical reaction force from the inclined tension field band. Where the flange is subjected compression, the girder can be assumed to be supported in the lateral direction due to the lateral/torsional buckling risk. The same steel quality as before, i.e. $f_y = 355$ MPa. The design bending moment of resistance for the class 4 effective cross-section can be assumed to meet the need of the load effect.

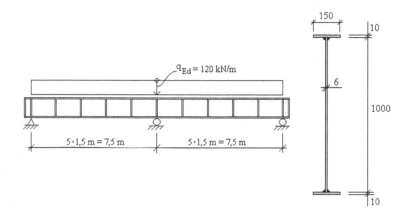

Message

The design shear panels here are also where the combined effect of maximum shear and maximum bending moment is the highest, in this case over the inner support. In relation to the girder in example 11, the increase in distance between the vertical stiffeners and the reduced flange dimension will influence the shear buckling resistance.

$$\frac{a}{b_w} = \frac{1.5}{1.0} = \underline{\underline{1.5}} \quad \Rightarrow \quad k_\tau = 5.34 + \frac{4.0}{1.5^2} = \underline{\underline{7.12}}$$

$$\tau_{cr} = k_\tau \cdot \frac{\pi^2 \cdot E}{12 \cdot (1 - v^2) \cdot \left(\dfrac{b_w}{t}\right)^2}$$

$$= 7.12 \cdot \frac{\pi^2 \cdot 2.1 \cdot 10^5}{12 \cdot (1 - 0.3^2) \cdot \left(\dfrac{1000}{6}\right)^2} = \underline{\underline{48.6 \text{ MPa}}}$$

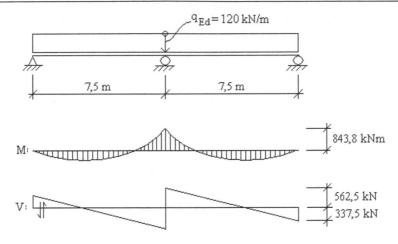

$$\overline{\lambda}_w = 0.76 \cdot \sqrt{\frac{f_{yw}}{\tau_{cr}}} = 0.76 \cdot \sqrt{\frac{355}{48.6}} = 2.05$$

$\overline{\lambda}_w > 1.08$ and intermediate support ("rigid end post"):

$$\Rightarrow \quad \chi_w = \frac{1.37}{(0.7 + \overline{\lambda}_w)} = \frac{1.37}{(0.7 + 2.05)} = 0.498$$

$$M_{f \cdot Rd} = \frac{A_f \cdot h_f \cdot f_{yf}}{\gamma_{M0}}$$

$$= \frac{150 \cdot 10 \cdot 1010 \cdot 10^{-9} \cdot 355 \cdot 10^3}{1.0} = 537.8 \, \text{kNm}$$

As the flange only design plastic moment of resistance do not exceed the design moment there is no contribution to the design shear resistance coming from the flanges (i.e. $V_{bf,Rd} = 0$):

$$V_{b,Rd} = V_{bw,Rd} = \frac{0.498 \cdot 355 \cdot 10^3 \cdot 1000 \cdot 6 \cdot 10^{-6}}{\sqrt{3} \cdot 1.0} = 612.4 \, \text{kN}$$

$$> V_{Ed} = 562.5 \, \text{kN} \quad \text{OK!}$$

A reduced shear buckling resistance in comparison to the girder in example 11 – due to the smaller flanges and the larger distance in between the vertical stiffeners – but still enough to meet the need. However, we also have to satisfy the interaction criterion for the combined effect of bending and shear. This criterion has only to be verified in the section located at a distance of $h_w/2(=0.5\,\text{m})$ from the inner support:

$x = 7.0\,\text{m}$:

$$V_{Ed} = 562.5 - 120 \cdot 0.5 = \underline{\underline{502.5\,\text{kN}}}$$

$$M_{Ed} = \frac{120 \cdot 7.0^2}{2} - 337.5 \cdot 7.0 = \underline{\underline{577.5\,\text{kNm}}}$$

$$W_{pl} = \frac{6 \cdot 1000^2}{4} + 2 \cdot 10 \cdot 150 \cdot 505 = \underline{\underline{3.015 \cdot 10^6\,\text{mm}^3}}$$

$$M_{pl,Rd} = \frac{3.015 \cdot 10^{-3} \cdot 355 \cdot 10^3}{1.0} = \underline{\underline{1070.3\,\text{kNm}}}$$

$$\frac{577.5}{1070.3} + \left(1 - \frac{537.8}{1070.3}\right) \cdot \left(2 \cdot \frac{502.5}{612.4} - 1\right)^2 = \underline{\underline{0.744}} \quad <1.0 \quad \text{OK!}$$

Literature

Behaviour and Design of Steel Plated Structures. ECCS (European Convention for Constructional Steelwork) TC8, Publication No 44, 1986.

ESDEP Lectures. European Steel Design Education Program. The Steel Construction Institute, Berkshire, England.

Ulfvarson, A.: **Buckling och knäckning.** (Local and global buckling. In Swedish). Chalmers University of Technology, Naval Architecture and Ocean Engineering, Göteborg, 1995.

Bresler, B. – Lin, T. Y. – Scalzi, J.: **Design of Steel Structures.** John Wiley & Sons, Inc., 1968.

McGuire, W.: **Steel Structures.** Prentice-Hall, Inc., 1968.

Galambos, T. (editor): **Guide to Stability Design Criteria for Metal Structures.** John Wiley & Sons, 1988.

Steel box girder bridges. Proceedings of the International Conference organized by the Institution of Civil Engineers in London, 13–14 February, 1973. The Institution of Civil Engineers, London, 1973.

Höglund, T.: **Dimensionering av stålkonstruktioner – utdrag ur Handboken Bygg, kapitel K18 och K19.** (Design of steel structures – extracts from the handbook Bygg (Engineering Design), chapter K18 and K19. In Swedish). C E Fritzes AB, Stockholm, 1994.

Samuelson, Å., m.fl.: **Skalhandboken – dimensionering mot buckling.** (The shell handbook – buckling design. In Swedish). Mekanförbundets förlag, Stockholm, 1990.

Ryall, J. M.: **Britannia Bridge, North Wales: Concept, Analysis, Design and Construction.** Proceedings of the International Historic Bridges Conference, Columbus, Ohio, USA, August 27–29, 1992, page 1–41.

Maquoi, R. – Skaloud, M.: **Stability of plates and plated structures. General Report.** Journal of Constructional Steel Research 55 (2000), page 45–68.

Bergfelt, A.: **Undersökning av bärförmågan hos lådbalkar av slanka plåtar, speciellt med avseende på instabilitetsfenomen.** (The investigation of the load-carrying capacity of slender plates, especially with respect to instability phenomena. In Swedish). Chalmers University of Technology, Department of Structural Engineering, Division of Steel and Timber Structures, Int.skr. S 78:7, Göteborg, 1978.

Report of Royal Commission into the Failure of West Gate Bridge. The West Gate Bridge Royal Commission Act 1970, No. 7989, Victoria, 1971. VPRS No. 2591/P0, Unit 14, Transport of Proceedings.

Aurell, T. – Englöv, P.: **Olyckor vid montering av stora lådbroar.** (Construction failures of large box-girder bridges. In Swedish). Väg- och vattenbyggaren 5, 1973, page 486–489.

Edlund, B.: **Katastrofen vid West Gate Bridge, Melbourne – ett dystert treårsminne.** (The catastrophe at the West Gate Bridge – a sad three-year memory. In Swedish). Väg- och vattenbyggaren 5, 1973, page 490–491.

Ekardt, H-P.: **Die Stauseebrücke Zeulenroda. Ein Schadensfall und seine Lehren für die Idee der Ingenieurverantwortung.** Stahlbau 67 (1998), Heft 9, page 735–749.

Edlund, B. – Crocetti, R.: **Bucklingsproblem vid stålbalkar under brolansering.** (Buckling problems in steel girders during launching. In Swedish). Chalmers University of Technology, Department of Structural Engineering and Mechanics, Steel and Timber Structures, Final Report to SBUF, Göteborg, 2003.

Crocetti, R.: **Design of plate girders – with special reference to EC3.** Chalmers University of Technology, Department of Structural Engineering, Steel and Timber Structures, Göteborg, 2001.

Knäckning, vippning och buckling – kommentarer till Stålbyggnadsnorm 70 (StBK-K2). (Global buckling, lateral/torsional buckling, and local buckling – comments to the Steel Design Code 70 (StBK-K2). In Swedish). Statens stålbyggnadskommitté. Svensk Byggtjänst, Stockholm, 1973.

Eurocode 3: Design of steel structures – Part 1–1: General rules and rules for buildings. SS-ENV 1993-1-1, Byggstandardiseringen, BST, Stockholm, 1995.

Eurocode 3: Design of steel structures – Part 1–1: General rules and rules for buildings. EN 1993-1-1: 2005.

Eurocode 3: Design of steel structures – Part 1–5: General rules: Supplementary rules for planar plated structures without transverse loading. ENV 1993-1-5: 1997.

Eurocode 3: Design of steel structures – Part 1–5: Plated structural elements. EN 1993-1-5: 2006.

Tjörnbron. (The Tjörn Bridge. In Swedish). Vägverket (Sune Brodin), Adlink, 1984.

Werner, E.: **Die Britannia- und Conway-Röhrenbrücke.** Werner-Verlag, Düsseldorf, 1969.

Hopkins, H. J.: **A Span of Bridges – an illustrated history.** Praeger Publishers, Inc., New York, 1970.

Åkesson, B.: **Buckling – ett instabilitetsfenomen att räkna med.** (Buckling – an instability phenomenon to reckon with. In Swedish). Studentlitteratur, Lund, 2005.

Picture and photo references

Figure 3.3   Ryall, J.M.: **Britannia Bridge, North Wales: Concept, Analysis, Design and Construction**. Proceedings of the International Historic Bridge Conference, Columbus, Ohio, USA, August 27–29, 1992, page 26.

Figure 3.4   Hopkins, H.J.: **A Span of Bridges – an illustrated history**. Praeger Publishers, Inc., New York, 1970, page 133.

Figure 3.7   Werner, E.: **Die Britannia- und Conway-Röhren-brücke**. Werner-Verlag, Düsseldorf, 1969, page 29.

Figure 3.10   **Conway Castle** from Gryffin Hill (post-card).

Figure 3.11   The **Britannia Bridge** as it looks like today after the fire in 1970 (photo taken by the author in September 1988).

Figure 3.17   Chalmers University of Technology, Department of Structural Engineering, Steel and Timber Structures, Göteborg (slide picture of the Fourth Danube Bridge in Wien).

Figure 3.22   Aurell, T. – Englöv, P.: **Olyckor vid montering av stora lådbroar**. (Construction failures of large box-girder bridges. In Swedish). Väg- och vattenbyggaren 5, 1973, page 487.

Figure 3.23   Aurell, T. – Englöv, P.: **Olyckor vid montering av stora lådbroar**. (Construction failures of large box-girder bridges. In Swedish). Väg- och vattenbyggaren 5, 1973, page 487.

Figure 3.30   Aurell, T. – Englöv, P.: **Olyckor vid montering av stora lådbroar**. (Construction failures of large box-girder bridges. In Swedish). Väg- och vattenbyggaren 5, 1973, page 488.

Figure 3.43   **Report of Royal Commission into the Failure of West Gate Bridge**. The West Gate Bridge Royal Commission Act 1970, No. 7989, Victoria, 1971, page 138. VPRS No. 2591/P0, Unit 14, Transport of Proceedings.

Figure 3.44   **Report of Royal Commission into the Failure of West Gate Bridge**. The West Gate Bridge Royal Commission Act 1970, No. 7989, Victoria, 1971, page 140. VPRS No. 2591/P0, Unit 14, Transport of Proceedings.

Figure 3.53   **Report of Royal Commission into the Failure of West Gate Bridge**. The West Gate Bridge Royal Commission Act 1970, No. 7989, Victoria, 1971, page 140. VPRS No. 2591/P0, Unit 14, Transport of Proceedings.

Figure 3.55   **Report of Royal Commission into the Failure of West Gate Bridge**. The West Gate Bridge Royal Commission Act 1970, No. 7989, Victoria, 1971, page 137. VPRS No. 2591/P0, Unit 14, Transport of Proceedings.

Figure 3.56 **Report of Royal Commission into the Failure of West Gate Bridge.** The West Gate Bridge Royal Commission Act 1970, No. 7989, Victoria, 1971, page 142. VPRS No. 2591/P0, Unit 14, Transport of Proceedings.

Figure 3.57 Some removed and saved parts from the collapsed **West Gate Bridge** at the Dep. of Civil Engineering, Monash University in Melbourne (with kind permission of prof. Em. Bo Edlund, Chalmers University of Technology, Göteborg, Sweden).

Figure 3.61 Chalmers University of Technology, Department of Structural Engineering, Steel and Timber Structures, Göteborg (slide picture of the **Rhine Bridge** in Koblenz).

Figure 3.70 Ekardt, H-P.: **Die Stauseebrücke Zeulenroda. Ein Schadensfall und seine Lehren für die Idee der Ingenieurverantwortung.** Stahlbau 67 (1998), Heft 9, page 738. With kind permission of Karl-Eugen Kurrer, Editor-in-chief, Stahlbau.

Figure 3.72 Ekardt, H-P.: **Die Stauseebrücke Zeulenroda. Ein Schadensfall und seine Lehren für die Idee der Ingenieurverantwortung.** Stahlbau 67 (1998), Heft 9, page 739. With kind permission of Karl-Eugen Kurrer, Editor-in-chief, Stahlbau.

Figure 3.73 Ekardt, H-P.: **Die Stauseebrücke Zeulenroda. Ein Schadensfall und seine Lehren für die Idee der Ingenieurverantwortung.** Stahlbau 67 (1998), Heft 9, page 736. With kind permission of Karl-Eugen Kurrer, Editor-in-chief, Stahlbau.

Figure 4.24 With kind permission of Alejandro Palomar, Mexico.

Figure 4.27 With kind permission of Alejandro Palomar, Mexico.

Figure 4.28 McGuire, W.: **Steel Structures.** Prentice-Hall, Inc., 1968, page 689.

Figure 4.31 McGuire, W.: **Steel Structures.** Prentice-Hall, Inc., 1968, page 689.

Figure 4.47 McGuire, W.: **Steel Structures.** Prentice-Hall, Inc., 1968, page 761.

Figure 5.2 **"Horror vacui".** Järnvägsteknik, No 2–3, 1976, page 59. With kind permission of Harald Axelsson, Statsbanornas Ingenjörsförening.

Figure 5.3 **Saltash Bridge** (post-card).

Figure 5.4 **Forth Bridge** (post-card).

Figure 5.16 **Globen.** Ny Teknik, No 11B, 1988.

Figure 5.19 The giant tanker **"Thorshammer"** of 228,259 tons passing the Almö Bridge (post-card).

Figure 5.20 The **Almö Bridge** (The Old Tjörn Bridge). Tjörn-bron. Vägverket (Sune Brodin), Adlink, 1984, page 6.

Index